π CHAPMAN & HALL/CRC
Monographs and Surveys in
Pure and Applied Mathematics   129

ISOMETRIES

ON BANACH SPACES:

function spaces

T0266502

CHAPMAN & HALL/CRC
Monographs and Surveys in Pure and Applied Mathematics

Main Editors

H. Brezis, *Université de Paris*
R.G. Douglas, *Texas A&M University*
A. Jeffrey, *University of Newcastle upon Tyne (Founding Editor)*

Editorial Board

R. Aris, *University of Minnesota*
G.I. Barenblatt, *University of California at Berkeley*
H. Begehr, *Freie Universität Berlin*
P. Bullen, *University of British Columbia*
R.J. Elliott, *University of Alberta*
R.P. Gilbert, *University of Delaware*
R. Glowinski, *University of Houston*
D. Jerison, *Massachusetts Institute of Technology*
K. Kirchgässner, *Universität Stuttgart*
B. Lawson, *State University of New York*
B. Moodie, *University of Alberta*
L.E. Payne, *Cornell University*
D.B. Pearson, *University of Hull*
G.F. Roach, *University of Strathclyde*
I. Stakgold, *University of Delaware*
W.A. Strauss, *Brown University*
J. van der Hoek, *University of Adelaide*

CHAPMAN & HALL/CRC
Monographs and Surveys in
Pure and Applied Mathematics   **129**

ISOMETRIES

ON BANACH SPACES:

function spaces

RICHARD J. FLEMING
JAMES E. JAMISON

CRC Press
Taylor & Francis Group
Boca Raton London New York

CRC Press is an imprint of the
Taylor & Francis Group, an **informa** business

A CHAPMAN & HALL BOOK

CRC Press
Taylor & Francis Group
6000 Broken Sound Parkway NW, Suite 300
Boca Raton, FL 33487-2742

First issued in paperback 2019

© 2003 by Taylor & Francis Group, LLC
CRC Press is an imprint of Taylor & Francis Group, an Informa business

No claim to original U.S. Government works

ISBN-13: 978-1-58488-040-0 (hbk)
ISBN-13: 978-0-367-39557-5 (pbk)
Library of Congress Card Number 2002041118

This book contains information obtained from authentic and highly regarded sources. Reasonable efforts have been made to publish reliable data and information, but the author and publisher cannot assume responsibility for the validity of all materials or the consequences of their use. The authors and publishers have attempted to trace the copyright holders of all material reproduced in this publication and apologize to copyright holders if permission to publish in this form has not been obtained. If any copyright material has not been acknowledged please write and let us know so we may rectify in any future reprint.

Except as permitted under U.S. Copyright Law, no part of this book may be reprinted, reproduced, transmitted, or utilized in any form by any electronic, mechanical, or other means, now known or hereafter invented, including photocopying, microfilming, and recording, or in any information storage or retrieval system, without written permission from the publishers.

For permission to photocopy or use material electronically from this work, please access www.copyright.com (http://www.copyright.com/) or contact the Copyright Clearance Center, Inc. (CCC), 222 Rosewood Drive, Danvers, MA 01923, 978-750-8400. CCC is a not-for-profit organization that provides licenses and registration for a variety of users. For organizations that have been granted a photocopy license by the CCC, a separate system of payment has been arranged.

Trademark Notice: Product or corporate names may be trademarks or registered trademarks, and are used only for identification and explanation without intent to infringe.

Library of Congress Cataloging-in-Publication Data

Fleming, Richard J.
    Isometries on Banach spaces : function spaces / by Richard J. Fleming and James E. Jamison.
        p. cm. — (Chapman & Hall/CRC monographs and surveys in pure and applied
    mathematics ; 129)
    Includes bibliographical references and index.
    ISBN 1-58488-040-6 (alk. paper)
    1. Function spaces. 2. Banach spaces. 3. Isometrics (Mathematics) I. Jamison, James E.
II. Title. III. Series.

QA323 .F55 2002
515′.73—dc21
                                               2002041118
                                                CIP

Visit the Taylor & Francis Web site at
http://www.taylorandfrancis.com

and the CRC Press Web site at
http://www.crcpress.com

Contents

Preface

Herman Weyl has said that in order to understand any mathematical structure, one should investigate its group of symmetries. In the class of Banach spaces, such a goal leads naturally to a study of isometries. A principal theme in geometry from the earliest times has been the study of transformations preserving lengths and angles. If the origins of the theory of Banach spaces are assigned to the appearance of Banach's book in 1932, then the study of Banach space isometries must be assigned the same starting date. In his book, Banach included the first characterizations of all isometries on certain classical spaces. The body of literature concerning isometries that has grown up since that time is large, perhaps surprisingly so.

An isometry, of course, is a transformation which preserves the distance between elements of a space. When Banach showed in 1932 that every linear isometry on the space of continuous real valued functions on a compact metric space must transform a continuous function $x(t)$ into a continuous function $y(t)$ satisfying

$$y(t) = h(t)x(\varphi(t))$$

where $|h(t)| \equiv 1$ and φ is a homeomorphism, he was establishing a canonical form characterization which fits in an astonishing number of cases. In this volume we are interested primarily in just such explicit descriptions of isometries. Our approach, then, will differ from that of many authors whose interest in isometries lies in showing which spaces are isometric to each other, or to those whose interest is in discovering properties of topological spaces on which the functions in the Banach spaces are defined. Such interests as these have served as excellent motivations for the types of characterizations that interest us. There have been several excellent surveys concerning isometries, including those by Behrends, Loomis, Jarosz, and Jarosz and Pathak, each of which concentrates on some subset of the whole. Our intent was to provide a survey of the entire subject, and our survey article (1993) serves as an inspiration and guide for the current work.

Our goal has been to produce a useful resource for experts in the field as well as beginners, and also for those who simply want to acquaint themselves with this portion of Banach space theory. We have tried to provide some history of the subject, some of the important results, some flavor of the wide variety of methods used in attacking the characterization problem for various types of spaces, and an exhaustive bibliography. We have probably underestimated the enormity of such a project, and perhaps what we have produced is more of a sampler than a full-blown survey.

We have chosen to organize the material according to the different classes of spaces under study and this is reflected in the chapter headings. The current volume is the first of two intended volumes, and as can be seen from the table of contents, it is primarily concerned with isometries on function spaces. The first chapter treats some general topics such as linearity, orthogonality, and Wold decompositions, while Chapter 6 contains material on noncommutative C^*-algebras, but the rest of the chapters treat the classical function spaces.

The second volume will include chapters with the following titles:

Chapter 7: The Banach-Stone Property
Chapter 8: Other Vector-Valued Function Spaces
Chapter 9: Orthogonal Decompositions
Chapter 10: Matrix Spaces
Chapter 11: Norm Ideals of Operators
Chapter 12: Spaces with Trivial Isometries
Chapter 13: Epilogue

In each chapter we try to include an early result of some historical importance. Other selections are made in order to expose some one of the principal methods that have been used, or perhaps to give an account of work that has not received much attention. The chapters, and even the sections within the chapters, are mostly independent of each other, and the reader can begin at any point of interest. In making our selections we have, no doubt, left out many others just as deserving and quite possibly more important. Hopefully, most of these omissions in the text are mentioned in the notes and the bibliography.

The exposition relies mostly on the original papers, and we have tried to report faithfully on those results, with additional clarifications when possible. Probably we have included more detail than necessary in some instances, but we have chosen to err on that side. There are a few places where we have given only sketchy arguments.

In each chapter we include a section on notes and remarks which give related results and other approaches that were not included in the main text. We hope these sections will help to soothe those who disagree sharply with our choice of material. For the most part, all references are given in the notes section. The exceptions are in cases where a reference is needed to justify a statement being made in the text.

In the bibliography we give a representative selection of the works on isometries, and a serious investigation of all such works available should probably begin with *Math Sci Net* using the phrases "isometries on," "isometries in," or "isometries for." Certainly there are many papers on vector-valued function spaces and matrix spaces which one might be expected to be mentioned here, but we are saving these for the second volume. That second volume will include many more references which were not directly relevant in this first one, and we also intend for Chapter 13 of Volume 2 to provide a further guide to the literature.

We assume that our readers are familiar with the standard material in courses in real variables, complex function theory, and functional analysis. Terms and notation that are common in those fields we leave undefined in the text. Page references to some of the special notation are given in the index. Some notation, of course, serves multiple purposes which should be understood in the context in which it appears. If some symbol or term is encountered which is not referenced in the index, the reader should be able to find it explained within a page or two of that location.

We have received much encouragement for this project from a number of people over recent years, and we want to mention three people in particular who have provided special help. Joe Diestel read portions of the work in early stages and his kind words helped move us forward. Bill Hornor has provided valuable advice, particularly in regard to Chapter 4 on analytic functions. David Blecher helped immeasurably in reading much of Chapter 6, and tried to guide us in understanding the material on operator spaces and the nonsurjective case of Kadison's theorem. However, we strongly emphasize the fact that we alone are responsible for any existing errors.

Finally, we would both like to express our deep appreciation and love for our wives, Diane Fleming and Jan Jamison, for their patience and devotion.

Richard J. Fleming and James E. Jamison

August 30, 2002

CHAPTER 1

Beginnings

1.1. Introduction

Isometries are, in the most general sense, transformations which preserve distance between elements. Such transformations are basic in the study of geometry which is concerned with rigid motions and properties preserved by them. The isometries of the Euclidean plane may all be described as rotations, translations, reflections, and glide reflections, and these transformations form a group under the operation of composition. This group is sometimes called the *Euclidean group* of the plane. Of course the Euclidean group is very large and often certain subgroups are sought which preserve some particular subset of the plane. If S is a subset of the Euclidean plane $\mathbb{R}^2$, the subgroup G which consists of all isometries which map S onto itself is called the *complete symmetry group* of S. A subgroup of G is called a *symmetry group* of S.

The symmetry group of the unit circle given by an equation

$$x_1^2 + x_2^2 = 1$$

relative to a fixed coordinate system is sometimes called the *orthogonal group* in the plane. It can be seen that each transformation in this group must leave the origin fixed and is therefore a linear transformation. This is a special case of a more general result which we will prove shortly.

The fact that transformations in the orthogonal group are linear allows them to be represented by matrices which are of two forms:

$$\begin{pmatrix} cos\theta & -sin\theta \\ sin\theta & cos\theta \end{pmatrix}$$

which represents a rotation through an angle θ; and

$$\begin{pmatrix} cos\theta & sin\theta \\ sin\theta & -cos\theta \end{pmatrix}$$

which represents a reflection of the plane with respect to the line given by the equation

$$x_1 sin(\theta/2) - x_2 cos(\theta/2) = 0.$$

The rotations form a subgroup of the symmetry group called the rotation group.

If instead of the circle, we find the symmetry group of a regular n-gon, we get the *dihedral group* D_n. For example, the case $n = 4$ gives the symmetry group for the square. Clearly rotations of $\pi/2$ and succeeding powers are isometries as are reflections across the four symmetry axes consisting of lines through opposite vertices and lines passing through midpoints of opposite sides. Thus D_4 has eight elements whose matrix forms are particularly nice. More generally, D_n consists of $2n$ elements which include n rotations in multiples of $2\pi/n$ and n reflections across symmetry axes as follows:

> If n is even, there are $n/2$ axes passing through opposite vertices and $n/2$ axes passing through midpoints of opposite sides; if n is odd, each edge has an opposite vertex and an axis passes through the midpoint of an edge and its opposite vertex.

We mention these well known facts simply to call attention to the basic role of the study of isometries in geometry as well as in elementary group theory. Before leaving this discussion of the isometries of the plane, however, let us note that the orthogonal group (that is, the symmetry group of the circle) is the set of all linear isometries of the real two dimensional Hilbert space $\ell^2(2)$. The dihedral group D_4 is the group of all linear isometries of the real two dimensional Banach space $\ell^\infty(2)$ (or for that matter, $\ell^p(2)$ for $1 \le p \le \infty$, $p \ne 2$). By $\ell^p(n)$ we mean the space of n-tuples of scalars with the p-norm given by $\|x\|_p = \left(\sum |x_j|^p\right)^{(1/p)}$ for $1 \le p < \infty$ and $\|x\|_\infty = \sup\{|x_j| : j = 1 \ldots n\}$. The group D_4 is very small in comparison with the orthogonal group which is a common observation when comparing isometries of a Banach space to those of a Hilbert space. In fact, a typical element of D_4 can be described by the simple formula

$$(1) \qquad\qquad g(t) = h(t)f(\varphi(t))$$

where $f = f(t)$ denotes a real function on $\{1, 2\}$, φ is a permutation of $\{1, 2\}$, and h is a real function with $|h(t)| = 1$ (unimodular). A formula something like (1) will appear often in the sequel because descriptions of isometries are usually given by this type of *canonical form*.

Herman Weyl [**321**, p.144] has said that in order to fully understand a mathematical structure one should investigate its group of symmetries. It is not surprising, then, that mathematicians would be interested in describing the isometries on particular Banach spaces and that this particular quest has led to a vast literature in the subject. Nor is it surprising to learn that the earliest results in the modern setting would be the work of Banach in his 1932 treatise on linear operators. Indeed, we open the door to our survey of the subject by examining in detail Banach's description of the form of an isometry on the space of continuous functions.

1.2. Banach's Characterization of Isometries on $C(Q)$

Suppose that Q is a compact metric space. By $C(Q)$ we mean the Banach space of continuous real valued functions defined on Q with the supremum norm. We begin with Banach's characterization of a peak point for a function f in relation to the existence of directional derivatives of the norm at f.

    1.2.1. LEMMA. *Let $f \in C(Q)$ and $s_0 \in Q$. In order that*

$$(2) \qquad |f(s_0)| > |f(s)| \quad \text{for each } s \in Q \text{ with } s \neq s_0,$$

it is necessary and sufficient that

$$(3) \qquad \lim_{t \to 0} \frac{\|f + tg\| - \|f\|}{t}$$

exists for each $g \in C(Q)$.

    Moreover, if f satisfies (2), we have

$$(4) \qquad \lim_{t \to 0} \frac{\|f + tg\| - \|f\|}{t} = g(s_0) sgn f(s_0)$$

for each $g \in C(Q)$.

    PROOF. Let us first show that (3) is necessary. If (2) holds, then $\|f\| = |f(s_0)|$. Now given $g \in C(Q)$ and a real number t, $f + tg$ is continuous on Q and so attains its maximum absolute value at some $s_t \in Q$. Therefore,

$$(5) \quad |f(s_0) + tg(s_0)| - |f(s_0)| \leq \|f + tg\| - \|f\| = |f(s_t) + tg(s_t)| - |f(s_0)|.$$

Also, we have

$$|f(s_0) + tg(s_0)| \leq |f(s_t) + tg(s_t)|$$

and a little manipulation yields the inequality

$$0 \leq |f(s_0)| - |f(s_t)| \leq |t||g(s_0)| + |t||g(s_t)| \leq 2|t|\|g\|.$$

It now follows that $\lim_{t \to 0} |f(s_t)| = |f(s_0)|$ and the compactness of Q allows us to conclude that

$$(6) \qquad \lim_{t \to 0} s_t = s_0.$$

    Now let us first suppose that $f(s_0) < 0$. By virtue of the fact that $s_t \to s_0$, we may choose t so small that

$$|f(s_0) + tg(s_0)| - |f(s_0)| = -f(s_0) - tg(s_0) + f(s_0) = -tg(s_0)$$

and

$$|f(s_t) + tg(s_t)| - |f(s_t)| = -f(s_t) - tg(s_t) + f(s_t) = -tg(s_t).$$

    From these two statements and (5) we see that

$$-tf(s_0) \leq \|f + tg\| - \|f\| \leq -tg(s_t)$$

for sufficiently small t which combined with (6) and the continuity of g leads to

$$\lim_{t \to 0} \frac{\|f + tg\| - \|f\|}{t} = -g(s_0).$$

The case where $f(s_0) > 0$ (which is the case considered in Banach's book) can be treated in a similar manner to establish that

$$\lim_{t \to 0} \frac{\|f + tg\| - \|f\|}{t} = g(s_0).$$

This completes the proof of the necessity of the existence of the limit in (3) and shows that (4) must hold.

For the sufficiency let us assume that $s_0, s_1 \in Q$ with $s_0 \neq s_1$ and

$$\|f\| = |f(s_0)| = |f(s_1)| \geq |f(s)|$$

for all $s \in Q$. If $f(s_0) < 0$, define $g(s) = -d(s, s_1)$ where d denotes the metric on Q. Then

$$\|f + tg\| - \|f\| \geq |f(s_0) + tg(s_0)| - |f(s_0)| = -f(s_0) - tg(s_0) + f(s_0)$$
$$= td(s_0, s_1)$$

for all sufficiently small t. We conclude that

(7) $$\lim_{t \to 0+} \inf \frac{\|f + tg\| - \|f\|}{t} \geq d(s_0, s_1) > 0.$$

However,

$$\|f + tg\| - \|f\| \geq |f(s_1) + td(s_1, s_1)| - |f(s_1)| = 0.$$

for all h, whereby we must have

(8) $$\lim_{t \to 0-} \sup \frac{\|f + tg\| - \|f\|}{t} \leq 0.$$

The inequalities (7) and (8) show that the limit (3) cannot exist.

For the case where $f(s_0) > 0$, we define $f(s) = d(s, s_1)$ and give a similar argument. □

We now state and prove the theorem of Banach for surjective isometries on $C(Q)$ spaces.

1.2.2. THEOREM. *If Q and K are compact metric spaces then for the spaces of real continuous functions $C(Q)$ and $C(K)$ to be isometrically isomorphic it is necessary and sufficient that Q and K be homeomorophic. In this case, an isometric isomorphism T from $C(Q)$ onto $C(K)$ must be given by*

(9) $$Tf(t) = h(t)f(\varphi(t)) \ \ for \ \ t \in K,$$

where φ is a homeomorphism from K onto Q and h is a real valued unimodular function on K.

PROOF. It is easy to see that if φ is a homeomorphism from K onto Q, then a transformation U defined by (9) is an isometric isomorphism of $C(Q)$ onto $C(K)$ and thus the sufficiency of the condition is clear.

For the necessity, let T be a linear isometry from $C(Q)$ onto $C(K)$, suppose $s_0 \in Q$ and let $f \in C(Q)$ be such that $|f(s_0)| > |f(s)|$ for all $s \in Q$. By Lemma 1.2.1,

$$\lim_{r \to 0} \frac{\|f + rg\| - \|f\|}{r} = g(s_0)sgnf(s_0)$$

must exist for every $g \in C(Q)$. Since T is an isometry,

$$(10) \qquad g(s_0)sgnf(s_0) = \lim_{r \to 0} \frac{\|f + rg\| - \|f\|}{r} = \lim_{r \to 0} \frac{\|Tf + rTg\| - \|Tf\|}{r}$$

and we apply the lemma again to conclude that there is some $t_0 \in K$ such that

$$|Tf(t_0)| > |Tf(t)| \quad for \ all \ t \in K \ with \ t \neq s_0.$$

(It is important to note here that we can apply the lemma since Tg runs through all of $C(K)$.) Furthermore, we may conclude from (10) and (4) that

$$g(s_0)sgnf(s_0) = Tg(t_0)sgnTf(t_0).$$

If we let $h(t_0) = sgnf(s_0)Tf(t_0)$, then $|h(t_0)| = 1$ and we get

$$(11) \qquad Ug(t_0) = h(t_0)g(s_0) \quad for \ each \ g \in C(Q).$$

Let us define ψ from Q to K by $\psi(s_0) = t_0$. Now ψ is injective, for if $\psi(s_1) = \psi(s_2)$, then by (11) we have $|g(s_1)| = |g(s_2)|$ for all $g \in C(Q)$ and so $s_1 = s_2$. To see that ψ is surjective, let $t_0 \in K$ and define q on K by

$$q(t) = \frac{1}{1 + d(t, t_0)}$$

where d denotes the metric on K. If $f = T^{-1}q$, then by (11)

$$|f(s)| = \frac{1}{1 + d(\psi(s), t_0)}$$

for each $s \in Q$. Since $\|f\| = \|q\| = 1$, there exists $s_0 \in Q$ such that $|f(s_0)| = 1$. Therefore, $\frac{1}{1 + d(\psi(s), t_0)} = 1$ which implies that $t_0 = \psi(s_0)$.

Finally, suppose that $\{s_n\}$ is a sequence in Q converging to s_0, $g \in C(K)$ and $Tf = g$. Since $|g(\psi(s_n)| = |f(s_n)|$ for each n and f is continuous, we have $|f(s_n)| \to |f(s_0)|$ so that $|g(\psi(s_n))| \to |g(\psi(s_0))|$ for every $g \in C(K)$. By choosing g defined by $g(t) = d(t, \psi(s_0))$, we obtain

$$d(\psi(s_n), \psi(s_0)) = |g(\psi(s_n))| \to |g(\psi(s_0))| = 0$$

from which we conclude that $\psi(s_n) \to \psi(s_0)$. This shows that ψ is continuous and since Q is compact and K is Hausdorff, ψ must in fact be a homeomorphism. Hence Q and K are homeomorphic and if we let $\varphi = \psi^{-1}$, we get from (11) the characterization of U given by (9).

$\square$

1.2.3. COROLLARY. *(Banach) If T is a surjective linear isometry on $C(Q)$ where Q is compact and metric, then*

$$Tf(t) = h(t)f(\varphi(t))$$

where $|h(t)| = 1$ and φ is a homeomorphism of Q onto itself.

We note that the statement in Corollary 1.2.3 gives the characterization of all surjective linear isometries on the given Banach space and it is this type of result that is the primary subject matter of this book. It is also important to see that Corollary 1.2.3 describes the symmetry group of the unit ball of the given space $C(Q)$, in the language given in the introduction, and that the elements of the symmetry group have the canonical form as given by (1). Of course, the space $\ell^\infty(2)$ discussed there is simply $C(Q)$ where Q is the set $\{1, 2\}$ given the discrete topology.

Banach's characterization then is truly the first attack on the general problem of identifying the symmetry group of the unit ball of a given Banach space; that is, characterizing the surjective, linear isometries on the space.

The proof of Banach's theorem uses in an essential way the assumptions that U is both linear and surjective. The surjective hypothesis will be invoked often (but not always) in remaining chapters, and the type of isometries of interest to us are the linear ones, so that we could have included the words "linear isometries" in the title. That the restriction to linear isometries is not really a serious defect is the subject of the next section.

1.3. The Mazur-Ulam Theorem

In our discussion of the isometries of the plane we noted that the orthogonal group consisted of linear transformations. Any linear transformation must fix the origin and any isometry U on a normed space X can be "shifted" so that it fixes the origin; just consider $V = U - U(0)$. It is a remarkable and useful fact, then, that any surjective isometry on a real normed linear space which fixes the origin must be linear.

To begin the discussion of the proof of this fact, let us introduce some notation and state a couple of lemmas. If x, y are elements of a n.l.s. X, let $H_1 = H_1(x, y)$ denote the set of elements $u \in X$ such that $\|x - u\| = \|y - u\| = \frac{1}{2}\|x - y\|$. For $n = 2, 3, \ldots$, let H_n be the set of $u \in H_{n-1}$ so that $\|u - v\| \leq \frac{1}{2}\delta(H_{n-1})$ for $v \in H_{n-1}$. Here, $\delta(H_{n-1})$ denotes the diameter of H_{n-1} which is, of course, the supremum of the distances between pairs of its elements. Clearly, $\delta(H_n) \leq \frac{1}{2^{n-1}}\|x - y\|$ for each n. Hence, the intersection of the H_n is either empty or consists of exactly one element which is called the (metric) *center* of the pair x, y.

1.3.1. LEMMA. *(Mazur-Ulam) If x, y are elements of a n.l.s. X, then $\frac{1}{2}(x + y)$ is the center of the pair x, y.*

PROOF. For each $u \in X$ let $\tilde{u} = x + y - u$. If $u \in H_1(x, y)$, then $\|\tilde{u} - x\| = \|y - u\|$ and $\|\tilde{u} - y\| = \|x - u\|$ so

$$\|\tilde{u} - x\| = \|\tilde{u} - y\| = \frac{1}{2}\|x - y\|$$

because $u \in H_1$. Assume that $\tilde{u} \in H_{n-1}$ whenever $u \in H_{n-1}$ and let $u \in H_n$. If $v \in H_{n-1}$, we have

$$\|\tilde{u} - v\| = \|(x + y - v) - u\| = \|\tilde{v} - u\| \le \frac{1}{2}\delta(H_{n-1})$$

since $\tilde{v} \in H_{n-1}$ and $u \in H_n$. Therefore, $\tilde{u} \in H_n$ as well and the upshot is that, by induction, for each positive integer n, $\tilde{u} \in H_n$ whenever $u \in H_n$.

Next we show by induction that $z = \frac{1}{2}(x + y) \in H_n$ for each n. First we see that $z \in H_1$ since $\|z - x\| = \frac{1}{2}\|x - y\|$. Assume that $z \in H_{n-1}$ and $u \in H_{n-1}$. Thus $\tilde{u} \in H_{n-1}$ by what we proved earlier and

$$2\|z - u\| = \|x + y - 2u\| = \|\tilde{u} - u\| \le \delta(H_{n-1}).$$

Hence, $\|z - u\| \le \frac{1}{2}\delta(H_{n-1})$ and we have $z \in H_n$. The conclusion is that $z \in \cap_1^\infty H_n$ and so is the center of x, y. □

If T is an isometry from a n.l.s. X into a n.l.s. Y, then T obviously maps $H_1(x, y)$ into $H_1(Tx, Ty)$, since

$$\|Tu - Tx\| = \|u - x\| = \frac{1}{2}\|x - y\| = \frac{1}{2}\|Tx - Ty\|$$

and similarly for $\|Tu - Ty\|$. This requires no linearity nor surjectivity for T.

1.3.2. LEMMA. *If T is a surjective isometry from X onto Y, then T maps the center of any pair x, y in X to the center of the pair Tx, Ty in Y.*

PROOF. Since T is a surjective isometry, it is clear that $T(H_1(x, y)) = H_1(Tx, Ty)$ and $\delta(H_1(x, y)) = \delta(H_1(Tx, Ty))$. If we assume this holds for $H_{n-1}(Tx, Ty)$, then for $u \in H_n(x, y)$ and $w \in H_{n-1}(Tx, Ty)$, we have $w = Tv$ for some $v \in H_{n-1}(x, y)$ while

$$\|Tu - Tv\| = \|u - v\| \le \frac{1}{2}\delta(H_{n-1}(x, y))\frac{1}{2}\delta(H_{n-1}(Tx, Ty)).$$

Therefore, $Tu \in H_n(Tx, Ty)$. Similarly, if $w \in H_n(Tx, Ty)$, then $w = Tu$ for some $u \in H_n(x, y)$. Thus by induction, $T(H_n(x, y)) = H_n(Tx, Ty)$ for every positive integer n. By Lemma 1.3.1, $\frac{1}{2}(x + y) \in H_n(x, y)$ for every n and it follows from the statement above that $T\left(\frac{1}{2}(x + y)\right) \in H_n(Tx, Ty)$ for every n. Hence $T\left(\frac{1}{2}(x + y)\right)$ is the unique element of $\cap_{n=1}^\infty H_n(Tx, Ty)$ which is $\frac{1}{2}(Tx + Ty)$ by Lemma 1.3.1 again. □

1.3.3. LEMMA. *If T is an isometry from a n.l.s. X onto a n.l.s. Y (real or complex), then*

(i) $T(x + y) = Tx + Ty - T(0)$,
(ii) $T(sx) = sTx + (1 - s)T(0)$ *for all real numbers s.*

PROOF. If $x \in X$, then from Lemma 1.3.2, we have

$$T(x) = T(\frac{1}{2}(2x + 0)) = \frac{1}{2}(T(2x) + T(0))$$

so that

(12) $$T(2x) = 2T(x) - T(0).$$

Now applying 1.3.2 again, $T(x + y) = T(\frac{1}{2}(2x + 2y)) = \frac{1}{2}(T(2x) + T(2y))$ and a double application of (12) yields (i) above.

Using (i) it is easy to show by induction that (ii) holds for all positive integers. Since

$$T(0) = T(2x + (-2x)) = T(2x) + T(-2x) - T(0)$$

we conclude that (ii) holds for $s = -1$ and therefore for any integer n. Upon applying (ii) to $T(x) = T\left(n\frac{x}{n}\right)$ we get that (ii) holds for all rationals and the extension to real follows from the continuity of T. $\qquad\square$

1.3.4. THEOREM. *If T is an isometry from a n.l.s. X onto a n.l.s. Y such that $\|Tx\| = 1$ for every $x \in X$ with $\|x\| = 1$, then $T(0) = 0$.*

PROOF. Suppose $z \in X$ and $Tz = 0$. Then

$$\|T0\| = \|Tz - T0\| = \|z - 0\| = \|z\|.$$

Suppose $z \neq 0$. By hypothesis and Lemma 1.3.3((ii)),

$$1 = \left\| T\left(\frac{1}{\|z\|}z\right) \right\| = \left|1 - \frac{1}{\|z\|}\right| \|z\|$$

from which it must be concluded that $\|z\| = 2$. However, we must also have

$$\left\| T\left(-\frac{1}{\|z\|}z\right) \right\| = \left|1 + \frac{1}{\|z\|}\right| \|z\|$$

which is absurd. Therefore $z = 0$, and the theorem is proved. $\qquad\square$

1.3.5. THEOREM. *(Mazur-Ulam) If T is an isometry from a n.l.s. X onto a n.l.s. Y, and if $T(0) = 0$, then T is real linear.*

PROOF. The proof is immediate from Lemma 1.3.3. $\qquad\square$

The assumption of surjectivity in the above theorem is necessary as can be seen from the following easy example.

1.3.6. EXAMPLE. *(Figiel) Let $X = \mathbb{R}$ with the absolute value norm and $Y = \ell^\infty(2, \mathbb{R})$. Then $U(a) = (a, \sin a)$ defines a nonlinear isometry from X into Y with $U(0) = (0,0)$.*

In order to remove the surjectivity assumption, we must put a condition on the range space. Recall that a normed linear space is *strictly convex* if $\|x + y\| = \|x\| + \|y\|$ implies that x, y are linearly dependent, and in fact, $x = ty$ for some $t \geq 0$.

1.3.7. LEMMA. *(Baker) If Y is a normed linear space which is strictly convex and $x, y \in Y$, then $H_1(x, y)$ is a singleton (and therefore consists of $\frac{1}{2}(x + y)$.*

PROOF. We have already seen that $\frac{1}{2}(x + y) \in H_1(x, y)$. Suppose $u, v \in H_1(x, y)$. Then

$$
(13) \qquad \left\| x - \frac{1}{2}(u + v) \right\| = \left\| \frac{1}{2}(x - u) + \frac{1}{2}(x - v) \right\|
$$

$$
\leq (1/2)\|x - u\| + (1/2)\|x - v\|
$$

$$
= \frac{1}{2}((1/2)\|x - y\|) + \frac{1}{2}((1/2)\|x - y\|)
$$

$$
= \frac{1}{2}\|x - y\|
$$

Similarly, $\|y - (1/2)(u + v)\| \leq \frac{1}{2}\|x - y\|$. If either inequality above is strict, then

$$
\|x - y\| \leq \|x - (1/2)(u + v)\| + \|y - (1/2)(u + v)\| < \|x - y\|.
$$

From this we can see that the inequalities must be equalities and

$$
\|(1/2)(x - u) + (1/2)(x - v)\| = \frac{1}{2}\|x - y\| = \|(1/2)(x - u)\| + \|(1/2)(x - v)\|.
$$

By the strict convexity, $(x - u) = t(x - v)$ for some $t \geq 0$ and since $\|x - u\| = \|x - v\|$, we have $t = 1$. It follows that $u = v$. $\qquad \square$

1.3.8. THEOREM. *(Baker) If T is an isometry from a real normed linear space X into a strictly convex, real normed linear space Y such that $T(0) = 0$, then T is linear.*

PROOF. For $x, y \in X$, we have seen that $(1/2)(x + y) \in H_1(x, y)$ and $T((1/2)(x + y)) \in H_1(Tx, Ty)$. However, $\frac{1}{2}(Tx + Ty) \in H_1(Tx, Ty)$ which is a singleton by Lemma 1.3.7. Hence,

$$
T((1/2)(x + y)) = \frac{1}{2}(Tx + Ty)
$$

and the proof is completed as in the proof of the Mazur-Ulam Theorem. $\qquad \square$

It can be seen that if Y is not strictly convex, then there is a normed linear space X and a nonlinear isometry T from X into Y such that $T(0) = 0$. In fact, we can take $X = \mathbb{R}$ and if u and v are linearly independent, norm-one vectors in Y with $\|u + v\| = \|u\| + \|v\|$, define T by

$$
T(t) = \begin{cases} tu, & \text{if } t \leq 1; \\ u + (t - 1)v, & \text{if } t > 1. \end{cases}
$$

The extension of these results to get complex linearity is not possible since the isometry $Uz = \bar{z}$ is not complex linear on $\mathbb{C}$. This simple operator is conjugate linear but the equally simple operator $T(z_1, z_2) = (z_1, \overline{z_2})$ is neither linear nor conjugate linear on $\mathbb{C}^2$.

1.4. Orthogonality

Isometries on Hilbert spaces preserve the inner product and they preserve orthogonality. Although these notions are not so natural in the Banach space setting, they can be defined and we want to see how they relate to isometries.

1.4.1. DEFINITION. *(Lumer) A semi-inner product (s.i.p.) on a complex vector space X is a complex valued form $[\cdot, \cdot]$ on $X \times X$ to $\mathbb{C}$ which satisfies*

(i) $[x, x] > 0$ *if* $x \neq 0$,
(ii) $[\alpha x + \beta y, z] = \alpha[x, z] + \beta[y, z]$ *for* $\alpha, \beta \in \mathbb{C}$, $x, y, z \in X$,
(iii) $|[x, y]|^2 \leq [x, x][y, y]$ *for all* $x, y \in X$.

It can be shown that $\|x\| = [x, x]^{1/2}$ defines a norm on X (with respect to which the s.i.p. is said to be compatible) and conversely, if $\| \cdot \|$ is a norm on X there is a s.i.p. on X which is compatible with the norm. This follows from the Hahn-Banach Theorem which guarantees the existence of duality maps $x \to \varphi_x$ which satisfy $\|\varphi_x\| = \|x\|$ and $\varphi_x(x) = \|x\|^2$, where $\varphi_x \in X^*$. The functional φ_x is called a *support functional*. Such a duality map gives a s.i.p. by means of the formula

$$[x, y] = \varphi_y(x).$$

Of course, such maps are not unique and so there may be many semi-inner products compatible with a given norm. It is always possible to choose a semi-inner product which satisfies

$$[s, \lambda y] = \overline{\lambda}[x, y].$$

1.4.2. DEFINITION. *(James) In a normed linear space $(X, \| \cdot \|)$, an element x is said to be* orthogonal *to y (written $x \perp y$) if*

$$\|x\| \leq \|x + \lambda y\| \quad \text{for all scalars } \lambda.$$

1.4.3. PROPOSITION. *(Giles) If $[\cdot, \cdot]$ is a semi-inner product compatible with the norm of X, then $[y, x] = 0$ implies that $x \perp y$.*

PROOF. Let λ be a given scalar. Then

$$\|x\|^2 = [x, x] = [x + \lambda y, x] \leq \|x + \lambda y\| \|x\|$$

so that $\|x\| \leq \|x + \lambda y\|$.   □

The converse of the proposition is not true. For example, if $x = (1, 1)$ and $y = (0, 1)$ in $\ell^\infty(2)$, then x is orthogonal to y but $[y, x] = 1/2$ where the semi-inner product is defined by

$$[y, x] = \begin{cases} y_1 \overline{x_1} & \text{if } |x_1| > |x_2|; \\ y_2 \overline{x_2} & \text{if } |x_1| < |x_2|; \\ \frac{1}{2} y_1 \overline{x_1} & \text{if } |x_1| = |x_2| \end{cases}$$

assuming $x = (x_1, x_2)$ and $y = (y_1, y_2)$.

We should note that in general if $x \perp y$, there is a s.i.p. compatible with the norm such that $[y, x] = 0$, but this choice of s.i.p. depends on the vectors

x, y. Thus we get a converse of sorts, and, in fact, the real thing when X is smooth.

1.4.4. PROPOSITION. *If x, y are elements of a n.l.s. X and $x \perp y$, then there is a s.i.p. $[\cdot, \cdot]$ compatible with the norm such that $[y, x] = 0$. If X is smooth, then $x \perp y$ if and only if $[y, x] = 0$ where $[\cdot, \cdot]$ is the unique s.i.p. compatible with the norm.*

PROOF. Suppose $x \perp y$. Then there is a linear functional f defined on the span of x and y such that $f(x) = \|x\|^2$ and $f(y) = 0$. For any scalars α, β,

$$|f(\alpha x + \beta y)| = |\alpha|\|x\|^2 \leq |\alpha|\|x\|\|x + \frac{\beta}{\alpha}y\| = \|x\|\|\alpha x + \beta y\|$$

Hence $\|f\| \leq \|x\|$ and it follows that $\|f\| = \|x\|$. By the Hahn-Banach Theorem there exists $x^* \in X^*$ with $\|x^*\| = \|f\| = \|x\|$, $x^*(y) = 0$ and $x^*(x) = \|x\|^2$. Let φ be a duality map such that $\varphi(x) = x^*$. Then the s.i.p. determined by φ satisfies the conclusion of the proposition.

If X is smooth, then there is only one duality map from X to X^*, hence only one s.i.p. compatible with the norm and it has the desired property. $\square$

1.4.5. THEOREM. *(Koehler and Rosenthal) Let X be a normed linear space and let U be a linear operator mapping X into itself. Then U is an isometry if and only if there is a semi-inner product $[\cdot, \cdot]$ such that $[Ux, Uy] = [x, y]$.*

PROOF. The sufficiency of the condition $[Ux, Uy] = [x, y]$ for U to be an isometry is trivial upon taking $x = y$.

On the other hand, if U is an isometry, let $[\cdot, \cdot]$ be any semi-inner product and note that for any $x, y \in X$,

$$[U^n x, U^n y] \leq \|U^n x\|\|U^n y\| = \|x\|\|y\|$$

so that $\{[U^n x, U^n y]\}$ is a bounded sequence of scalars. Let F be a linear functional of norm 1 on ℓ^∞ such that (i) $F((1, 1, \ldots)) = 1$ and (ii) $F((x_n)) = F((x_{n+1}))$ (i.e., F is a Banach limit whose existence is given by Banach [**21**, p.34]). Then define

$$[[x, y]] = F(\{[U^n x, U^n y]\}).$$

The properties of F guarantee that $[[x, y]]$ defines a semi-inner compatible with the norm and $[[Ux, Uy]] = [[x, y]]$ follows by property (ii) of F. $\square$

In a Hilbert space it is easy to see that a linear isometry must preserve orthogonality because it must preserve the inner product. Conversely, a non-zero linear operator T on a Hilbert space which preserves orthogonality must be a positive multiple of an isometry. To see that, suppose first that x is a nonzero element of a Hilbert space X with inner product $\langle \cdot, \cdot \rangle$ and let y be any nonzero element for which $Ty \neq 0$ and such that $\langle x, y \rangle \neq 0$. Let $\alpha \neq 0$

be such that $\langle x + \alpha y, y \rangle = 0$. If $Tx = 0$, and since T preserves orthogonality we get that

$$\alpha \langle Ty, Ty \rangle = 0$$

from which we conclude $\alpha = 0$ and $\langle x, y \rangle = 0$ contrary to assumption. Hence $Tx \neq 0$ whenever $x \neq 0$. Now suppose x, y are any two nonzero elements of X which are not orthogonal. Again choose α so that $\langle x + \alpha y, y \rangle = 0$ and therefore $\langle Tx + \alpha Ty, Ty \rangle = 0$ by the given property of T. The conclusion is that

$$\frac{\langle x, y \rangle}{\langle y, y \rangle} = -\alpha = \frac{\langle Tx, Ty \rangle}{\langle Ty, Ty \rangle}.$$

Thus

(14) $$\frac{\langle Ty, Ty \rangle}{\langle y, y \rangle} = \frac{\langle Tx, Ty \rangle}{\langle x, y \rangle}$$

and by interchanging the roles of x and y in (14) we get

$$\frac{\langle Tx, Tx \rangle}{\langle x, x \rangle} = \frac{\langle Ty, Tx \rangle}{\langle y, x \rangle} = \frac{\overline{\langle Tx, Ty \rangle}}{\overline{\langle x, y \rangle}} = \frac{\langle Ty, Ty \rangle}{\langle y, y \rangle},$$

or

(15) $$\frac{\langle Tx, Tx \rangle}{\langle x, x \rangle} = \frac{\langle Ty, Ty \rangle}{\langle y, y \rangle}.$$

If $\langle x, y \rangle = 0$, choose $z \neq 0$ which is orthogonal to neither and apply (14) to the pairs x, z and z, y to conclude that (15) holds for all pairs x, y which are not zero. From this, it follows that if we fix y and let

$$r = \left(\frac{\langle Ty, Ty \rangle}{\langle y, y \rangle} \right)^{1/2},$$

then $\|Tx\| = r\|x\|$ for all $x \in X$ so that $U = \dfrac{1}{r}T$ is the desired isometry.

This argument makes heavy use of the properties of inner products and does not carry over to the Banach space case. Yet there is a germ of an idea here which Koldobsky has exploited to obtain the result, at least for real Banach spaces.

1.4.6. THEOREM. *(Koldobsky) Let X be a real Banach space and T a linear operator on X to itself which preserves orthogonality. Then $T = rU$ where $r \in \mathbb{R}$ and U is an isometry.*

To explain the proof of this theorem we must first introduce some notation and observe some facts. If x and y are given elements of a normed linear space X, then continuity of the norm guarantees that there is at least one α such that $x + \alpha y$ is orthogonal to y since $\|x + ry\|$ must attain its minimum value for some scalar α. However, such α is not necessarily unique and the convexity of the function $\alpha \to \|x + \alpha y\|$ (and the continuity) implies that the set $A(x, y)$ of all α such that $(x + \alpha y) \perp y$ is a closed, bounded interval $[a, b]$, in the case where X is a real Banach space.

It is useful to note here that if T is a linear operator which preserves orthogonality, and $x \neq 0$, then $Tx \neq 0$. For if $Tx = 0$, one can find y with $Ty \neq 0$ and x not orthogonal to y. There exists $\alpha \neq 0$ such that y is orthogonal to $x + \alpha y$ and so Ty must be orthogonal to itself which is not possible.

For a nonzero x in X, let

$$D(x) = \{x^* \in X^* : \|x^*\| = 1, \ x^*(x) = \|x\|\}.$$

Then for every $y \in X$,

(16) $$\lim_{\alpha \to 0+} \frac{\|x + \alpha y\| - \|x\|}{\alpha} = sup\{x^*(y) : x^* \in D(x)\}$$

and the corresponding limit as $\alpha \to 0^-$ is given by

$$inf\{x^*(y) : x^* \in D(x)\}.$$

Hence, the function φ defined on $\mathbb{R}$ by $\varphi(\alpha) = \|x + \alpha y\|$ is differentiable at α if and only if $x_1^*(y) = x_2^*(y)$ for every $x_1^*, x_2^* \in D(x + \alpha y)$. Finally, we observe here that if x, y are linearly independent, the function φ is convex on $\mathbb{R}$ and so differentiable almost everywhere with respect to Lebesgue measure. Let $D_\varphi(x, y)$ denote the set of all α at which φ is differentiable; then $\mathbb{R} \setminus D_\varphi(x, y)$ is of measure zero.

It is convenient to state and prove some lemmas which are due to Koldobsky.

1.4.7. LEMMA. *If $\alpha \in D_\varphi(x, y)$ and $\beta, \gamma \in \mathbb{R}$, then*

(i) *$\|x + \alpha y\| = \|x + ay\|$ for all $\alpha \in A(x, y) = [a, b]$.*
(ii) *The number $x^*(\beta x + \gamma y)$ does not depend on $x^* \in D(x + \alpha y)$.*
(iii) *$x + \alpha y \perp \beta x + \gamma y$ if and only if $x^*(\beta x + \gamma y) = 0$ for every $x^* \in D(x + \alpha y)$.*
(iv) *Either $x + \alpha y \perp y$ or there exist a unique number $f(\alpha) \in \mathbb{R}$ such that $x + \alpha y \perp x - f(\alpha)y$.*

PROOF. (i) As we observed above, if $x + \alpha y \perp y$, then $\|x + \alpha y\|$ is the minimum value of $\|x + ry\|$ for $r \in \mathbb{R}$, hence is the same for all such α.

(ii) We have already seen that $x^*(y)$ is the same for all $x^* \in S(x + \alpha y)$. Since

$$x^*(x) = x^*(x + \alpha y) - \alpha x^*(y) = \|x + \alpha y\| - \alpha x^*(y),$$

we see that this number is also independent of $x^* \in S(x + \alpha y)$ and therefore so is the linear combination $\beta x + \gamma y$.

(iii) If $x + \alpha y \perp \beta x + \gamma y$, then by (16),

$$\sup_{x^* \in S(x+\alpha y)} x^*(\beta x + \gamma y) = \lim_{r \to 0+} \frac{(\|x + \alpha y + r(\beta x + \gamma y)\|)}{r} \geq 0$$

and similarly,

$$\inf_{x^* \in S(x+\alpha y)} x^*(\beta x + \gamma y) \leq 0.$$

It follows from ((ii)) above that

$$sup \; x^*(\beta x + \gamma y) = inf \; x^*(\beta x + \gamma y) = 0.$$

If $x^*(\beta x + \gamma y) = 0$ for every $x^* \in S(x + \alpha y)$, then

$$x^*(x + \alpha y + r(\beta x + \gamma y)) = x^*(x + \alpha y) = ||x + \alpha y||$$

for every r and since $||x^*|| = 1$, we have $x + \alpha y \perp \beta x + \gamma y$.

(iv) Let $x^* \in S(x + \alpha y)$. If $x^*(y) = 0$, then $x + \alpha y \perp y$ by (3) above. Otherwise, $x + \alpha y \perp x - \beta y$ if and only if $x^*(x - \beta y) = 0$ so we may choose $\beta = f(\alpha) = x^*(x)/x^*(y)$.   □

As previously observed, the set $A(x, y)$ of all α such that $x + \alpha y \perp y$ is a closed interval which we denote by $[a, b]$ in the lemma below.

1.4.8. LEMMA.
(i) *For every* $\alpha > b$, $||x + \alpha y|| = ||x + by|| exp \left(\int_b^\alpha (t + f(t))^{-1} dt \right)$.
(ii) *For every* $\alpha < a$,

$$||x + \alpha y|| = ||x + ay|| exp \left(- \int_\alpha^a (t + f(t))^{-1} dt \right).$$

PROOF. (i) If $\alpha \in D(x, y)$ with $\alpha > b$ and $x^* \in S(x + \alpha y)$, we have $x^*(x) = f(\alpha)x^*(y)$ and $x^*(y) = \varphi'(\alpha)$ where $\varphi(\alpha) = ||x + \alpha y||$. Then

$$x^*(x) = x^*(x + \alpha y) - \alpha x^*(y) = ||x + \alpha y|| - \alpha \varphi'(\alpha)$$

which leads to

$$\frac{\varphi'(\alpha)}{||x + \alpha y||} = (\alpha + f(\alpha))^{-1}.$$

Now the set $\mathbb{R} \backslash D(x, y)$ has Lebesgue measure zero, and we can write

$$\int_b^\alpha \frac{\varphi'(t)}{\varphi(t)} dt = \int_b^\alpha (t + f(t))^{-1} dt \quad \text{for every } \alpha > b.$$

The function $\alpha \rightarrow ln \; \varphi(\alpha)$ satisfies a Lipschitz condition and is absolutely continuous. Hence

$$ln \; \varphi(\alpha) - ln \; \varphi(b) = \int_b^\alpha \frac{\varphi'(t)}{\varphi(t)} dt = \int_b^\alpha (t + f(t))^{-1} dt$$

from which we obtain

$$||x + \alpha y|| = ||x + by|| exp \int_b^\alpha (t + f(t))^{-1} dt.$$

The proof of (ii) is similar.   □

We are now ready to give the proof of Theorem 1.4.6.

PROOF. We assume that T is a nonzero operator on X which preserves orthogonality. Let x be such that $Tx \neq 0$ and suppose y is any element of X so that x and y are linearly independent. Since T preserves orthogonality, it is clear that $A(x, y)$ is contained in $A(Tx, Ty)$. Suppose $\alpha \in A(Tx, Ty)$. If $\alpha \in A(x, y)$, then by Lemma 1.4.7(iv), there exists $f(\alpha)$ such that $x + \alpha y \perp$

$x - f(\alpha)y$ and so $Tx + \alpha Ty \perp Tx - f(\alpha)Ty$. If $x^* \in S(Tx + \alpha Ty)$ we have by Lemma 1.4.7(iii) that $x^*(Ty) = 0$ and $x^*(Tx - f(\alpha)Ty) = 0$ so that we must conclude that

$$0 = x^*(Tx + \alpha Ty) = \|Tx + \alpha y\|.$$

This is not possible since $x + \alpha y$ is not orthogonal to y and therefore $T(x + \alpha y)$ cannot be zero. The contradiction requires us to conclude that $A(x, y) = A(Tx, Ty)$ so that the numbers a, b and the function $f(\alpha)$ are the same for both pairs (x, y) and (Tx, Ty).

As we know from Lemma 1.4.7 (i), there are constants c_1, c_2 such that $\|x + \alpha y\| = c_1$ and $\|Tx + \alpha Ty\| = c_2$ for $\alpha \in [a, b]$. It now follows from Lemma 1.4.8 that $\|Tx + \alpha Ty\| = \frac{c_2}{c_1}\|x + \alpha y\|$ for every $\alpha \in \mathbb{R}$. For $\alpha = 0$ we get $\|Tx\| = \frac{c_2}{c_1}\|x\|$; for $\alpha > 0$ we have $\|\frac{Tx}{\alpha} + Ty\| = \frac{c_2}{c_1}\|\frac{x}{\alpha} + y\|$, and as $\alpha \to \infty$ we get $\|Ty\| = \frac{c_2}{c_1}\|y\|$. Hence, $\frac{\|Tx\|}{\|x\|} = \frac{\|Ty\|}{\|y\|}$ for every $x, y \neq 0$ and the conclusion of the theorem follows.

$\square$

1.5. The Wold Decomposition

A linear isometry on a Hilbert space which is surjective is called a unitary operator and this term is sometimes used to refer to surjective isometries on general Banach spaces. It is a well known fact that every isometry on a Hilbert space can be written as a direct sum of a unitary operator and copies of the unilateral shift. A proof of this result makes use of the fact that if V is an isometry on H, then $M = \cap V^n H$ reduces V and the restriction $V|M$ is unitary. If L is the orthogonal complement of $V(H)$, then V restricted to the closed linear span N of the subspaces $V^n(L)$ is a direct sum of copies of the unilateral shift and $H = M \oplus N$. The multiplicity of the shift is the dimension of L.

This decomposition, sometimes referred to as the *Wold Decomposition*, lays bare the structure of all isometries on Hilbert spaces. Of course, the unitary operators are numerous and completely determined by pairings of orthonormal bases, while the structure of unilateral shift operators is also well understood. In the Banach space setting, isometries which are not surjective often present difficulties, and if the goal is to find a crisp description such as the canonical form, the search is often restricted to the surjective case.

In some instances, a type of Wold Decomposition can be obtained and we want to say a little about that here. The given development is the work of Campbell, Faulkner, and Sine.

1.5.1. DEFINITION. *Let V be an injective linear map on a Banach space X. Then V is said to be a* unilateral shift *provided there is a subspace L of X for which $X = \oplus_0^\infty V^n(L)$. That is, X is the direct sum of the spaces $V^n(L)$. The subspace L is referred to as a* wandering subspace *and the dimension of L is called the* multiplicity of V.

1.5.2. DEFINITION. *An isometry V of a Banach space X is said to be a Wold isometry provided $X = M_\infty \oplus N_\infty$ where $M_\infty = \cap_1^\infty V^n(X)$ and $N_\infty = \sum_0^\infty \oplus V^n(L)$. Here, the space L is a complement for the range of V, and N_∞ is written as a Schauder decomposition. (This means that each element x has a unique representation as $x = \sum_0^\infty x_n$ where $x_n \in V^n(L)$.) Note that V restricted to M_∞ is a surjective isometry while V restricted to N_∞ is a shift.*

Suppose V is an isometry on X and $V(X)$ is complemented in X. Let L denote a complement of $V(X)$ so that $X = V(X) \oplus L$, and suppose P_1 is a bounded projection from X onto $V(X)$. Let $P_2 = V P_1 V^{-1} P_1$ and if P_n has been defined, let $P_{n+1} = V P_n V^{-1} P_n$. The inductively defined sequence $\{P_n\}$ is an abelian family of projections with P_k projecting X onto $V^k(X)$ and annihilating $L \oplus V(L) \oplus V^2(L) \oplus \cdots \oplus V^{k-1}(L)$. This sequence will be referred to as a *sequence of projections associated with V*.

1.5.3. THEOREM. *(Campbell, Faulkner, and Sine) Let V be an isometry of a reflexive Banach space. If the range of V is complemented and V has a uniformly bounded associated sequence of projections, then V is a Wold isometry.*

PROOF. Let X denote a reflexive Banach space and V and isometry on X as described in the hypotheses. Let L and $\{P_n\}$ be as described above. Then we are assuming that there is some real number $b > 0$ so that $\|P_n\| \leq b$ for every n. Let $M_\infty = \cap_1^\infty V^n(X)$ and $N_\infty = \oplus_0^\infty V^n(L)$ where L denotes the complement of $V(X)$ and the null space of P_1.

If $x \in X$, then $x = P_n x + (I - P_n)x$ for each n and $\{P_n x\}$, $\{(I - P_n)x\}$ form bounded sequences. By the Eberlein-Smulian Theorem [89, p.430], there exist x_0, y_0 which are weak limits of a subsequence of $\{P_n x\}$ and $\{(I - P_n)x\}$ respectively so that $x = x_0 + y_0$. For a given n, $P_k(x) \in V^n(X)$ for all k sufficiently large and $V^n(X)$ is weakly closed so that $x_0 \in V^n(X)$. Therefore $x_0 \in M_\infty$ and since $(I - P_n)x \in N_\infty$ for each n, $y_0 \in N_\infty$.

Next we see that if $x \in M_\infty \cap N_\infty$, then we can choose a sequence $\{n_k\}$ of integers with the property that $x_{n_k} \in L + V(L) + \cdots + V^{n_k-1}(L)$ and $x_{n_k} \to x$. It follows that $P_{n_k}(x - x_{n_k}) = x$ and $\|P_n(x - x_{n_k})\| \leq b\|x_{n_k} - x\|$. Hence $x = 0$ and we conclude that $X = M_\infty \oplus N_\infty$. The final piece is to see that N_∞ can be written as a Schauder decomposition $\sum_0^\infty \oplus V^n(L)$.

If $y \in N_\infty$, then y can be approximated by elements from $L + V(L) + \cdots + V^{n-1}(L)$ and P_n annihilates such elements. From this we see that $P_n(y) \to 0$ and since $(I - P_n)(y) = \sum_0^{n-1}(P_k(y) - P_{k+1}(y))$ we may conclude that $y = \sum_0^\infty (P_k(y) - P_{k+1}(y))$ where $P_k(y) - P_{k+1}(y) \in V^k(L)$ for each k. This expansion is unique because if $z \in V^i(L) \cap V^j(L)$ with $i < j$, then $0 = P_j(z) = z$.   □

An isometry whose range is 1-complemented is said to be *orthocomplemented*. The language reflects the fact that there is a projection P from X

onto $V(X)$ of norm 1, if and only if $V(X) \perp (I-P)(X)$ in the sense discussed in Section 4. Hence, in this case, for every $x, y \in X$, $V(y) \perp (x - Px)$.

1.5.4. COROLLARY. *If V is an orthocomplemented isometry on a reflexive Banach space, then V is a Wold isometry.*

Of course not every isometry is complemented, and complemented isometries need not be orthocomplemented. However, in the case of the L^p spaces, something can be said.

1.5.5. COROLLARY. *Let (Ω, Σ, μ) be a σ-finite measure space. If V is an isometry on $L^p(\Omega, \Sigma, \mu)$ $(1 < p < \infty)$, then V is a Wold isometry.*

PROOF. By a result of Ando [9], every isometry on L^p is orthocomplemented. $\square$

We close this section with a pair of examples which exhibit Wold decompositions.

1.5.6. EXAMPLE. *Let $X = L^p[0, 1]$ where $1 < p < \infty$. Define $V : L^p \to L^p$ by $Vf(t) = f(\varphi(t))$ where*

$$\varphi(t) = \begin{cases} 2t, & \text{if } 0 \leq t \leq 1/2; \\ 2 - 2t & \text{if } 1/2 < t \leq 1. \end{cases}$$

Then V is a linear isometry which is not surjective.

It is straightforward to verify that V is indeed an isometry. However, g is in the range of V if and only if $g(t) = g(1 - t)$ for all $t \in [0, 1]$. Hence V is not surjective and we want to describe its Wold Decomposition.

One interesting way to describe this isometry is in terms of its action on the Walsh functions which form a complete orthonormal system in L^p. Recall that the *Rademacher functions* $\{r_n(t)\}$ are defined by

$$r_n(t) = sign \ (sin \ 2^n \pi t) \ \text{for } t \in [0, 1].$$

Thus $r_1(t)$ is 1 on $(0, 1/2)$ and -1 on $(1/2, 1)$, while $r_2(t)$ is 1 on $(0, 1/4)$ and $(1/2, 3/4)$ and -1 on $(1/4, 1/2)$ and $(3/4, 1)$. The pattern for the rest should be clear. We let $r_0(t) \equiv 1$. The *Walsh functions* w_k are given by $w_1 = r_0$ and $w_{k+1}(t) = r_{n_1+1}(t) r_{n_2+1}(t) \ldots r_{n_N+1}(t)$ where $k = 2^{n_1} + 2^{n_2} + \cdots + 2^{n_N}$ for integers $n_1 > n_2 > \ldots n_N \geq 0$. It will be helpful later to denote the number N above as $d(k)$.

Now our isometry V takes the product of two functions to the product of their images and it is not too hard to see that $Vr_n = r_{n+1}r_1$ for every $n \geq 1$. These two properties, together with the fact that an even power of any Rademacher function is identically one, make it possible to establish the following lemma.

1.5.7. LEMMA.
(i) *If $k \geq 1$ and $d(k)$ is odd, then $V(w_{k+1}) = w_{2(k+1)}$.*
(ii) *If $k \geq 1$ and $d(k)$ is odd, then w_{k+1} is not in $V(X)$.*

(iii) *If $k \geq 1$ and $d(k)$ is even, then $V(w_{k+1}) = w_{2k+1}$.*

(iv) *If $k \geq 1$ and $d(k)$ is even, then*

$$w_{k+1} = \begin{cases} Vw_{(k+2)/2} & \text{if } k \text{ is even;} \\ Vw_{(k+1)/2} & \text{if } k \text{ is odd.} \end{cases}$$

Let us now give the proof that V has a Wold Decomposition.

PROOF. If we let

$$L = \overline{sp}\{w_{k+1} : d(k)\text{is odd}\} = \overline{sp}\{w_2, w_3, w_5, w_8, w_9, w_{12}, \ldots\},$$

then L is a complement for $V(X)$. The product of an even number of Rademacher functions is a Walsh function which is in $V(X)$ and the product of two elements in $V(X)$ must also be in $V(X)$. The unique semi-inner product on L^p which is compatible with the norm is given by

$$[f, g] = \int_0^1 fg \frac{|f|^{p-2}}{\|g\|^{p-2}}.$$

(We are assuming the real case here.)

If r_n is any Rademacher function, then $r_n(t) = -r_n(1-t)$ and if $f \in V(X)$, then

$$[r_n, f] = \int_0^1 r_n(t)f(t)\frac{|f(t)|^{p-2}}{\|f\|^{p-2}}dt$$
$$= \int_0^{1/2} r_n(t)f(t)\frac{|f(t)|^{p-2}}{\|f\|^{p-2}} + \int_{1/2}^1 r_n(t)f(t)\frac{|f(t)|^{p-2}}{\|f\|^{p-2}}$$
$$= \int_0^{1/2} r_n(t)f(t)\frac{|f(t)|^{p-2}}{\|f\|^{p-2}}dt + \int_0^{1/2} r_n(1-s)f(1-s)\frac{|f(1-s)|^{p-2}}{\|f\|^{p-2}}ds$$
$$= 0$$

so that $f \perp r_n$ by Proposition 1.4.3. If w_{k+1} is a Walsh function with $d(k)$ odd, then $w_{k+1} = wr_n$ where w is a product of an even number of Rademacher functions. Then for $f \in V(X)$, $wf \in V(X)$ and $[w_{k+1}, f] = [r_n, wf] = 0$ so that $f \perp w_{k+1}$. We conclude that $V(X) \perp L$ and so by the remark preceding Corollary 1.5.4, there is a projection P from X onto $V(X)$ along L of norm 1.

The corollary says that V is a Wold isometry. It is easy to see by induction that $V^n r_k = r_{n+k}r_n$ for each n and k from which one can conclude that the only Walsh function in $M_\infty = \cap_1^\infty V^n(X)$ is w_1. Thus $M_\infty = sp\{w_1\}$, $N_\infty = L \oplus V(L) \oplus \ldots$, and V splits into the identity on M_∞ and a shift of infinite multiplicity on N_∞. $\square$

1.5.8. EXAMPLE. *Let X be any of the spaces ℓ^∞, (c), (c_0). Define V on X by $V(\alpha_1, \alpha_2, \ldots) = (\alpha_1, \alpha_1, \alpha_2, \ldots)$. Then V is an isometry and $V(X)$ is all those sequences $x = (\alpha_j)$ for which $\alpha_1 = \alpha_2$. Thus V is a Wold isometry in case $X = c_0$ or $X = c$, but not when $X = \ell^\infty$.*

PROOF. If $L = sp\{(1, 0, 0, \ldots)\}$, then $X = V(X) \oplus L$ and $P(\alpha_1, \alpha_2, \ldots) = (\alpha_2, \alpha_2, \alpha_3, \ldots)$ is a norm one projection from X onto $V(X)$ along L. From the definition of V it is clear that any element in $\cap V^n(X)$ must be a constant sequence. If $X = (c_0)$ then $M_\infty = \{0\}$ and $(c_0) = L \oplus V(L) \oplus \cdots$ so that V is a pure shift (which shifts a basis for (c_0)). If $X = (c)$, then $M_\infty = sp\{(1, 1, 1, \ldots)\}$ and $N_\infty = L \oplus V(L) \oplus \cdots = (c_0)$. Here V is the sum of the identity on M_∞ and a basis shift on N_∞. Finally, if $X = \ell^\infty$, we have a breakdown since M_∞ is still just the constant sequences while $M_\infty \oplus (L \oplus V(L) \oplus \cdots)$ does not give all of ℓ^∞. $\square$

1.6. Notes and Remarks

According to J.L. Coolidge [79, p.273], the earliest proof of the characterizations of "transformations which keep distances invariant in the plane as rotations, translations, or a product of one of these and a reflection in a line" was given by Chasles [66]. However, Coolidge believes it must have been known much earlier since the three dimensional version was studied by Euler [96]. Although we have suggested that Banach was the first to describe isometries in the sense that is the focus of this current work, Banach himself [21, p.241] says that, "The oldest known example of an isometry between spaces of type B is the one, between L^2 and ℓ^2, which was obtained by Riesz-Fischer and Parseval-Fatou." In fact, the first characterization of isometries on a particular space may have been by I. Schur [281] in connection with matrix spaces.

The group point of view has been considered by a number of authors. Most closely related to our discussion in the introduction is a paper by Putnam and Wintner [259] which studies the orthogonal group in Hilbert space. Another interesting paper by Stern [294] characterizes groups which are isomorphic to the group of linear isometries on some Banach space.

We want to mention here some references in which the author or authors give a survey of the characterization of isometries in some setting. Included are books by Behrends [23] and Jarosz [147], a paper of Jarosz and Pathak [151], and the Ph.D. thesis of Loomis [201]. There is also the book of Lacey [191] which is concerned with isometric theory, but the purpose and focus of that book is somewhat different from that of the current volume.

Banach's characterization of isometries on C(Q). Theorem 1.2.2 (or Corollary 1.2.3) is the original version of what has since been called the Banach-Stone Theorem. Stone proved the Theorem in the case where the spaces Q and K are assumed only to be compact and Hausdorff [295]. If $C(Q, E)$ denotes the space of continuous functions from a compact, Hausdorff space Q into a Banach space E, then E is said to have the *Banach-Stone Property* if the conclusions of Theorem 1.2.2 hold for isometric isomorphisms from $C(Q, E)$ to $C(K, E)$. Thus Stone showed that the real numbers have the Banach-Stone Property. Jerison [153] was the first to consider the question for more general Banach spaces E. These issues are discussed fully in later chapters.

Theorem 1.2.2 and/or its Corollary 1.2.3 have been proved in a number of ways and by all of the principal methods that are used to find isometries in Banach spaces. Banach's method of using the directional derivative of the norm to identify peak points of functions has not appeared often in the works of other authors. Sundaresan [298] has proved a generalization of Lemma 1.2.1 and used it in a proof of a Banach-Stone Theorem for spaces $C(Q, E)$, but not in a direct way to find the homeomorphism as Banach did. Okikiolu [240] used differentiation of norms to determine isometries in L^p-spaces. W. Werner [320] used a differentiability property to characterize isometries on C^*-algebras. This will be treated in some detail in Chapter 6.

Eilenberg [92] discusses Banach's proof, showing that it is necessary to know that each point of Q is a G_δ set. Eilenberg actually obtains the result under the assumption that Q and K are completely regular and satisfy the first axiom of countability. The reference to Theorem 1.2.2 as the Banach-Stone Theorem in Eilenberg's paper may be the first time it was so named.

The fact that Corollary 1.2.3 actually describes the symmetry group of the unit sphere for $C(Q)$ follows from Theorems 1.3.4 and 1.3.5 which guarantee that a symmetry of the surface of the unit ball in any normed linear space must be a linear isometry.

Mazur-Ulam Theorem. Theorem 1.3.5 was proved by Mazur and Ulam [219] in response to a question raised by Banach. Our organization is a bit different, but the key result is Lemma 1.3.1. The proof we have given for that lemma is essentially that given by Banach in his book [21, p.166]. We have added some details by means of Lemma 1.3.2 and Lemma 1.3.3. We have stated Theorem 1.3.4 in order to emphasize the connection with the notion of symmetry groups and the fact that preservation of even the surface of the unit sphere by an isometry (i.e., a symmetry of the sphere) guarantees it must be linear.

Banach gave the name *rotation* to an isometry which fixes a point, so by the Mazur-Ulam Theorem, every rotation, indeed every isometry, is a translate of a linear isometry.

The following conjecture appears in Mankiewicz [210] but the question was first raised by Banach [21, p.241].

1.6.1. CONJECTURE. *If (X, d) is a linear metric space, then every isometry T from (X, d) onto an arbitrary linear metric space (Y, d') with $T(0) = 0$ is linear.*

Thus Mazur and Ulam have established the conjecture when X and Y are normed linear spaces. Charzynski [65] showed that (1.6.1) holds when (X, d) is finite dimensional, and Mankiewicz [209], [210] showed that the conjecture is true if (X, d) is a locally convex Montel space or if X is locally convex and satisfies the strong Krein-Milman property.

Charzynski's proof is considered to be very difficult [209], [268] but he did show that the conjecture must hold in the general case if it holds when $X = Y$. Wobst [322] has given a simpler proof of Charzynski's theorem.

Rolewicz [268, Theorem 9.3.12] established the conjecture (1.6.1) under the assumption that X and Y are locally bounded spaces with the properties that the functions $d(tx, 0)$ on X and $d'(ty, 0)$ on Y are concave for each t. His proof follows the lines of Mazur-Ulam but is adjusted to reflect the lack of homogeneity in the "F-norms" $d(x, 0)$, $d'(x, 0)$.

Ratz [265] has generalized the Mazur-Ulam Theorem in another direction, to vector spaces furnished with a nondegenerate symmetric bilinear form.

Lemma 1.3.7, Theorem 1.3.8 and the example following the proof of the theorem are due to J.H. Baker [20], whose paper is often overlooked in other discussions of the non-surjective case. Theorem 1.3.8 or close relatives are mentioned in Edelstein [90], Bosznay [43], Wells and Williams [319, Lemma 12.1], and Drewnowski [88]. In this latter paper Drewnowski, states the following interesting question:

1.6.2. QUESTION. *Let X and Y be Banach spaces such that there exists an isometric embedding of X into Y. Does there also exist a linear isometric embedding of X into Y?*

He shows that if T is an isometry from a Banach space X into $C_0(Q)$, where Q is locally compact, then there is a closed subset K of Q and a linear isometry T_K from X into $C_0(K)$ where $T_K(x) = Tx|K$. This generalizes a result due to Holsztynski [135] and Lovblom [202]. Drewnowski's paper contains a number of interesting results about the question of linearity. There is also a good discussion of the Mazur-Ulam Theorem in Day [83].

Example 1.3.6 is due to Figiel [99] who proved the following theorem which he says was conjectured by Holsztynski and Lindenstrauss.

1.6.3. THEOREM. *(Figiel) Let X and Y be two real Banach spaces and U an isometry (not necessarily linear) mapping X into Y such that $U(0) = 0$. If the linear span of $U(X)$ is dense in Y, then there is a continuous linear operator F mapping Y into X such that the composition $F \circ U$ is the identity on X. The operator F is uniquely determined and has norm one.*

In a paper that appears back-to-back with that of Figiel, Holsztynski [134] proves the very same theorem using ideas from category theory and a theorem of Kaplansky [167] concerning order isomorphisms.

In the original paper, Mazur and Ulam [219] simply state that each surjective isometry which takes zero to zero must be additive and so their proof does not use the assumption that the spaces were real. As the examples show, it is impossible to always get linearity in the complex case. It is tempting to hope that one might get some kind of combination of linearity and conjugate linearity, but examples of complex spaces not isomorphic to their complex conjugates (Bourgain [44] and Kalton [160]) suggest that one proceed with caution.

Orthogonality. The notion of semi-inner product was given by Lumer [203] who wanted to be able to bring Hilbert space type arguments to bear in the Banach space setting. His use of this notion in his pioneering paper [205] on the isometries of Orlicz spaces opened up a new and useful method of finding isometries which is often referred to as Lumer's Method. It involves the notion of Hermitian operators on Banach spaces and the fact that UTU^{-1} must be Hermitian if T is Hermitian and U is an invertible isometry. One way of describing a Hermitian operator is by use of the s.i.p.; T is *Hermitian* if $[Tx, x]$ is real for all x where $[\cdot, \cdot]$ is compatible with the norm. Then if U is an isometry, the fact that UTU^{-1} must be Hermitian is a consequence of Theorem 1.4.5 of Koehler and Rosenthal [172]. The reader should consult both Lumer [203] and Giles [115] for general information about semi-inner products. The part of Proposition 1.4.4 involving smoothness is really Theorem 2 in [115]. We will have considerable to say about Lumer's method in a later chapter.

In addition to Theorem 1.4.5 which we have included, Koehler and Rosenthal [172] investigate some spectral properties for isometries. For example, they show that eigenvectors corresponding to distinct eigenvalues of an isometry are mutually orthogonal.

The paper of R.C. James [143] is probably still the best source of general information about orthogonality in Banach spaces. Working with this notion in Banach spaces can be treacherous business and Koldobsky's proof [179] that operators which preserve it must be isometries is an impressive effort. We ended up including nearly everything in his paper in order to make the proof understandable. Thus Theorem 1.4.6 and Lemmas 1.4.7 and 1.4.8 are taken from [179]. It is not known whether the result holds for complex spaces.

The Wold Decomposition. Our discussion of the Wold Decomposition of a Hilbert space isometry is based on the development given by Halmos [122]. The result in its earliest form is due to Von Neumann [313] and in a statistical setting by Wold [323]. The earliest references to a "Wold Decomposition" seem to be in papers of Kolmogorov [182] (see [283]) and Doob [86]. Subsequent discussions have appeared in a number of places such as [8], [87], [212], and [232]. Anderson's discussion [8] may be the one to read for those who want to see how the statistical version and Hilbert space version compare.

The extension to the Banach space case first appeared in Faulkner and Honeycutt [98]. This was followed by the paper of Campbell, Faulkner, and Sine [58] which presented more powerful generalizations and a sharper analysis of the pure and unitary factors. We have lifted three results (Theorem 1.5.3, and Corollaries 1.5.4 and 1.5.5) from the latter paper. Some interesting examples are given there as well as a look at the question of when an isometry shifts a basis. An interesting treatment of this type of question about basis shifting can be found in a paper of Gutek, Hart, Jamison, and Rajagopalan [121].

The description of the Rademacher and Walsh functions used in our Example 1.5.6 comes from Singer [288], [289] but of course these are mentioned in many places. In fact, the Walsh functions form a Schauder basis for L^p ($1 < p < \infty$), a result mentioned in [289, p.572] and attributed to Paley [241].

We will see that in some cases, nonsurjective isometries can be described in the "canonical" sense given earlier, and the reader can look for these in the work of Forelli [106], Holsztynski [133], Lamperti [193], Novinger [237], and others.

CHAPTER 2

Continuous Function Spaces-The Banach-Stone Theorem

2.1. Introduction

In Chapter 1 we gave Banach's original proof of the characterization of linear isometries from $C(Q, \mathbb{R})$ to $C(K, \mathbb{R})$ where K and Q are compact metric spaces. Stone improved the result in 1937 by giving the same characterization (Theorem 1.2.2) where K and Q are compact Hausdorff spaces. Although Stone's proof is not a main focus in this chapter, it is of such historical importance that we cannot resist giving at least a sketch of it here.

2.1.1. THEOREM. *(Stone) If K and Q are compact Hausdorff spaces and T is an isometric isomorphism of $C(Q, \mathbb{R})$ onto $C(K, \mathbb{R})$, then there is a homeomorphism φ from K onto Q and continuous unimodular function h on K such that for each $f \in C(Q)$,*

(17) $$Tf(t) = h(t)f(\varphi(t)) \ \text{ for } t \in K.$$

PROOF. (Sketch) We will content ourselves with indicating how Stone produced the h and the φ.

For $s \in Q$, let $Q(s) = \{f \in C(Q) : |f(s)| = \|f\|\}$. Then $Q(s)$ contains all the real constant functions and if $f_1, f_2, \ldots, f_n$ are in $Q(s)$ then for

$$g = \sum_{j=1}^{n} f_j \, sgn f_j,$$

we see that $g \in Q(s)$ and $\|g\| = \sum \|f_j\|$.

If s is any fixed element in Q, let $F(f)$ denote the set of all t in K such that $|Tf(t)| = \|Tf\|$ where $f \in Q(s)$. Given $f_1, f_2, \ldots f_n$ in $Q(s)$ and $g = \sum f_j sgn f_j$ as above, there must exist a $t \in K$ such that $|Tg(t)| = \|Tg\|$. However,

$$\|Tg\| = |Tg(t)| \leq \sum |Tf_j(t)| \leq \sum \|Tf_j\| = \sum \|f_j\| = \|g\| = \|Tg\|$$

and we conclude that $|Tf_j(t)| = \|UTf_j\|$ for each $j = 1, \ldots, n$. Therefore, the closed sets $F(f)$ (for $f \in Q(s)$) have the finite intersection property and since K is compact there is at least one $t \in K$ common to all. Thus $Tf \in K(t)$ for every $f \in Q(s)$, and U maps $Q(s)$ into $K(t)$ for some t. In the same way, T^{-1} maps $K(t)$ into $Q(r)$ for some r. In fact, we must have $s = r$ and so there is a one-to-one map $s \to t = \psi(s)$ which establishes a one-to-one correspondance between the sets $Q(s)$ and $K(t)$. The sets $\{s \in Q : |f(s)| = \|f\|\}$ correspond to $\{t \in K : |Tf(t)| = \|Tf\|\}$ as do their complements which are shown to form a basis for the topologies of Q and K, respectively.

If we let $\varphi = \psi^{-1}$ and $h = U1$, then (17) follows after certain manipulations, which we omit.

□

As noted in Chapter 1, this theorem, soon to be known as the Banach-Stone Theorem, has been proved by many authors and by varied methods. We will try to relate some of that story in the notes at the end of this chapter. In this chapter itself, we wish to examine in some detail three approaches to the theorem which supply generalizations of different kinds and which have not, to our knowledge, been discussed very much in other sources. The first is due to Eilenberg, whose effort was the earliest after Banach and Stone, and which generalizes slightly the type of topological spaces Q and K. The second is due to Holsztynski and Novinger and treats linear isometries of subspaces of $C(Q)$ into $C(K)$. Finally, we present a more recent result of Vesentini, who obtains the canonical form (17) for bounded transformations from $C(Q)$ into $C(K)$ which take extreme points to extreme points.

In this chapter we will not discuss the important vector-valued case but reserve that to a later chapter on the *Banach-Stone Property*.

2.2. Eilenberg's Theorem

As we have seen in both Banach's proof and Stone's proof, the desired homeomorphism arises by identifying points where given functions achieve their norms. In Banach's proof, the identification of so-called peak functions involves the existence of the directional derivative of the norm function. Eilenberg suggested another way to identify these functions.

Let us say that a function f in the space $C_b(K)$ of bounded continuous functions on a topological space K is a *peak function* with $t_0 \in K$ as *peak* if

$$|f(t)| < |f(t_0)| \text{ for all } t \in K \text{ with } t \neq t_0.$$

2.2.1. DEFINITION. *If X is a Banach space and $x \in X$ with $\|x\| \leq 1$, then the* star *of x (denoted by $st(x)$) is the set of all $y \in X$ with $\|y\| \leq 1$ such that $\|x + y\| = 2$.*

Clearly, $x \in st(x)$ if and only if $\|x\| = 1$ and in fact any element of a star set must have norm exactly one.

2.2.2. LEMMA. *(Eilenberg) Suppose K is countably compact with $f, g \in C(K)$ and $\|f\| \leq 1$, $\|g\| \leq 1$ (the functions may be either real or complex valued). Then $f \in st(g)$ if and only if there is $t_0 \in K$ such that $f(t_0) = g(t_0)$ and*

$$|f(t_0)| = |g(t_0)| = 1.$$

PROOF. The sufficiency of the condition is obvious. For the necessity, the countable compactness guarantees $t_0 \in K$ so that

$$2 = \|f + g\| = |f(t_0) + g(t_0)|.$$

The result follows from some elementary complex arithmetic. ☐

The next theorem gives a simple characterization of peak functions.

2.2.3. THEOREM. *(Eilenberg) Let K be countably compact, completely regular, and $C(K)$ the space of continuous (real or complex) functions on K. A function $f \in C(K)$ with $\|f\| = 1$ is a peak function if and only if $st(f)$ is convex.*

PROOF. If f is a peak function, there exists $t_0 \in K$ such that $|f(t_0)| = 1$ and $|f(t)| < 1$ for $t \neq t_0$. From Lemma 2.2.2, it is clear that $st(f) = \{g \in C(K) : \|g\| = 1$ and $g(t_0) = f(t_0)\}$. This set is obviously convex.

Now suppose that f is not a peak function. Then $|f|$ must attain its maximum value 1 (by the countable compactness) and it must occur for at least two distinct points $t_1, t_2 \in K$. Since K is completely regular, there are disjoint open sets G_1, G_2 containing t_1, t_2, respectively, as well as continuous norm-one functions g_1, g_2 with $g_i(t_1) = f(t_i)$ and $g_i(t) = 0$ for $t \in K \backslash G_i$, $i = 1, 2$. Then

$$\|f + g_1\| \geq |f(t_1) + g_1(t_1)| = 2$$

so that $\|f + g_1\| = 2$; similarly $\|f + g_2\| = 2$. However, $\|g_1 + g_2\| = 1$ by the construction of g_1 and g_2 so that both are in $st(f)$ while $\frac{1}{2}(g_1 + g_2)$ is not. Therefore, $st(f)$ is not convex. ☐

We are now in a position to give the proof of Eilenberg's slight extension of Banach's 1932 result. The key idea, as in Banach's proof, is that an isometry must map peak functions to peak functions, and it is also crucial that for every $t_0 \in K$ there is a peak function which peaks at t_0. This latter statement is true if and only if K is completely regular and every singleton point of K is a G_δ-set. The characterization of peak functions given by Theorem 2.2.3 makes it easy to establish the desired homeomorphism. It does take some work to get the canonical form of the isometry and we are going to give some of those details. We are also going to give the proof for the complex case whereas Banach, Stone, and Eilenberg all treated the case of real-valued functions only.

2.2.4. THEOREM. *(Eilenberg) Suppose Q and K are completely regular, countably compact spaces satisfying the first axiom of countability. Let T be an isometric isomorphism of $C(Q, \mathbb{C})$ onto $C(K, \mathbb{C})$. Then there is a homeomorphism φ of K onto Q and a unimodular function $h \in C(K, \mathbb{C})$ such that*

(18) $$Tf(t) = h(t)f(\varphi(t)) \quad \text{for all } t \in K.$$

PROOF. Let $s_0 \in Q$. Since K satisfies the first axiom, there is a countable base $\{G_n\}$ of open neighborhoods of s_0 and since K is completely regular, $\{s_0\} = \cap G_n$. It follows that there is a peak function f which peaks at s_0. By Theorem 2.2.3, $st(f)$ is convex and since T is an isometry, $st(Tf) = T(st(f))$ is convex as well so that Tf is a peak function. Therefore, there is some $t_0 \in K$ such that Tf peaks at t_0.

It is important to note that if g is any norm-one function which attains its norm at s_0, then we can let $v = \dfrac{f(s_0)}{g(s_0)}g$ to get a function which is in $st(f)$. It follows that $Tv \in st(Tf)$ from which we conclude that $Tv(t_0) = Tf(t_0)$ where $|Tf(t_0)| = 1$ (Lemma 2.2.2). Hence $|Tg(t_0)| = 1$ as well which shows that the function ψ which pairs s_0 with t_0 is well defined. We will see that it is a homeomorphism.

If $\psi(s_1) = \psi(s_2)$ and $s_1 \neq s_2$, then there are peak functions f_1, f_2 which peak at s_1 and s_2, respectively. Since Tf_1 and Tf_2 both peak at $\psi(s_1) = \psi(s_2)$, application of T^{-1} would imply that f_1 and f_2 must both peak at s_1 which is not true. It is also straightforward to show that ψ is surjective, using the fact the T^{-1} is an isometry.

For the continuity, let $s_0 \in Q$ and suppose G is a neighborhood of $\psi(s_0)$. Since K is completely regular, we may choose $g \in C(K)$ with $\|g\| = 1$ such that $g(\psi(s_0)) = 0$ and $g(t) = 1$ for all $t \in K \backslash G$. Let $f \in C(Q)$ be such that $Uf = g$. Then $N = \{s \in Q : |f(s)| < \|f\|\}$ is an open neighborhood of s_0 and $\psi(N) \subseteq G$ since f and $Uf = g$ must peak at points which correspond under ψ.

It remains to show now that T satisfies (18). From what we have already shown, we must have $|T1(t)| = 1$ for every t. Let $h = T1$ and $V = \overline{h}T$. Then $V(1) = 1$, and we will show that if $v \in C(Q)$ with $v(s) = 0$, then $Vv(\psi(s)) = 0$. The canonical form (18) follows from this since for every $s \in Q$ and $f \in C(Q)$, $v = f - f(s)$ vanishes at s and so

$$0 = Vv(\psi(s)) = Vf(\psi(s)) - f(s).$$

If we write $\varphi(t) = \psi^{-1}(t)$ and replace V by $\overline{h}T$, we get (18).

Hence, to complete the proof we suppose first that v is a non-negative function in $C(Q)$ with $v(s_0) = 0$, $v(s) \neq 0$ for $s \neq s_0$, and $\|v\| = 1$. With these assumptions, $1 - v$ is a peak function which peaks at s_0 and therefore $V(1 - v) = 1 - V(v)$ is a peak function which peaks at $\psi(s_0)$. Now $1 - v$ is in $st(1)$ so that $1 - V(v) \in st(V1) = st(1)$. By Theorem 2.2.3, there is some $t_0 \in K$ at which $V(1 - v)(t_0) = 1$. However, $|V(1 - v)(\psi(s_0))| = 1$ and since $V(1 - v)$ is peak function, we must conclude that $t_0 = \psi(s_0)$. Thus

$$1 = V(1 - v)(\psi(s_0)) = 1 - V(v)(\psi(s_0))$$

and we have $Vv(\psi(s_0)) = 0$. If $\|v\| \neq 1$ we replace v by $\dfrac{v}{\|v\|}$ and get the same result.

If f is any non-negative function with $f(x_0) = 0$ and v is as above, then $v + f$ satisfies the hypotheses above so that $V(v + f)(\psi(s_0)) = 0$ and we get

$Vf(\psi(s_0)) = 0$. By the linearity of V we can extend the result, first to all real functions by using the decomposition into positive and negative parts, and finally to complex functions using the real and imaginary parts.   □

We close this section with the observation that Eilenberg was principally motivated by the desire to relate the structure of a topological space K to the metric structure of $C_b(K)$. In this connection he constructed from $C(K)$ an "ideal space" that he showed (in case K is compact) to be homeomorphic to K and so obtained an independent proof of Stone's Theorem. By making use of the Stone-Cech compactification, he actually extended the classical theorem to the case where Q and K are completely regular and satisfy the first axiom of countability.

2.3. The Nonsurjective Case

The success of the arguments of Banach, Stone, and Eilenberg depended in part on the assumption that the isometry T is surjective. Holstzynski was the first to relax that assumption without making any other assumptions about U. In this section, following the work of Novinger, we go a step further and describe isometries which map a subspace M of $C(Q)$ onto a subspace N of $C(K)$. The proof uses an extreme point argument first used by Arens and Kelley and which is perhaps the best known method of proof for the standard Banach-Stone Theorem as popularized by Dunford and Schwartz [89, p.441].

We will need a characterization of extreme points of the unit ball in the dual of a $C(K)$ space. Such a characterization was developed by Arens and Kelley and is known as the Arens-Kelley Theorem. We will have use of this result in several places and so we present a proof of it here, choosing a form of the theorem more general than needed, but which will also be useful in a later chapter.

If K is a locally compact Hausdorff space and E a normed linear space (real or complex), we let $C_0(K, E)$ denote the space of all continuous functions f from K to E which vanish at infinity with

$$\|f\| = \sup_{t \in K} \|f(t)\|_E.$$

We will denote the closed unit ball of any normed linear space E by the symbol $B(E)$, the surface of the closed ball by $S(E)$, and the set of extreme points of a given set D by $ext(D)$. In the particular case where the set D is the closed unit ball of a Banach space E, we will simply write $ext(E)$.

We are going to follow the method of Ströbele [297] in characterizing extreme points of the dual unit ball for certain special quotient spaces of $C_0(K, E)$. We begin by proving the Buck-Phelps characterization of extreme points in the dual ball of a general normed linear space.

We will use the common notation $\Re z$ and $\Im z$ for the real and imaginary parts of a complex number.

2.3.1. LEMMA. *(Buck and Phelps) Suppose E is a normed linear space (real or complex). Then z^* is an extreme point of $B(E^*)$ if and only if $E(z^*)-E(z^*) = E$ where $E(z^*) = \{x \in E : \|x\| - \Re(z^*(x)) \leq 1\}$.*

PROOF. Let $z^* \in B(E^*)$. The convex set $E(z^*) - E(z^*)$ has a nonempty interior and so if it is not all of E, then there exists y^* in E such that $\Re y^*(x) \leq 1$ for all $x \in E(z^*) - E(z^*)$ and in fact $|\Re y^*(x)| \leq 1$ for all such x. Let $x \in B(E)$ and suppose $\alpha = \|x\| - \Re z^*(x)$. First assume $\alpha = 0$. Then $\lambda x \in E(z^*)$ for every $\lambda > 0$ and since $|y^*(x)| \leq \lambda$ for all λ, we must have $y^*(x) = 0$. It follows that $\Re(z^* \pm y^*)(x) = \Re z^*(x) \leq 1$. If $\alpha > 0$, then $\alpha^{-1}x \in E(z^*)$ so that $|\Re y^*(\alpha^{-1}x)| \leq 1$. Therefore,

$$\pm \Re y^*(x) \leq \alpha = \|x\| - \Re z^*(x)$$

and $\Re(z^* \pm y^*)(x) \leq 1$. We must conclude that $z^* \pm y^* \in B(E^*)$ and so z^* cannot be an extreme point.

On the other hand, if z^* is not an extreme point of $B(E^*)$, there must be a nonzero $y^* \in E^*$ such that $z^* \pm y^* \in B(E^*)$. For $x \in E$, $\Re(y^*(x)) \leq \|x\| - \Re z^*(x)$ so that $\Re(y^*(x)) \leq 1$ for all $x \in E(z^*)$. Hence $\Re(y^*(x) \leq 2$ for all $x \in E(z^*) - E(z^*)$ and this set cannot be all of E.   □

2.3.2. LEMMA. *(Buck and Phelps) If M is a closed subspace of a n.l.s. E (real or complex), then z^* is an extreme point of $B((E/M)^*) = B(E^*) \cap M^\perp$ if and only if $E(z^*) - E(z^*) + M = E$.*

PROOF. By Lemma 2.3.1, z^* is an extreme point of $B((E/M)^*)$ if and only if

$$(19) \qquad (E/M)(z^*) - (E/M)(z^*) = E/M.$$

If $z \in E(z^*)$, where $z^* \in M^\perp$, then $[z] \in (E/M)(z^*)$ where $[z]$ denotes the equivalence class of E/M corresponding to z. It follows readily that if $E(z^*) - E(z^*) + M = E$, then (19) holds and z^* is extreme.

If we assume that z^* is extreme and $w \in E$, then $[w] = [z] - [u]$ for $[z], [u] \in (E/M)(z^*)$. Given $\epsilon > 0$, we may choose a real number λ such that $0 < \lambda < 1$ and

$$\|[w] - \lambda([z] - [u])\| < \epsilon.$$

Now

$$\|[\lambda z]\| - \Re z^*(\lambda z) \leq \lambda < 1$$

and

$$\|[\lambda u]\| - \Re z^*(\lambda u) \leq \lambda < 1,$$

and it is possible to choose $m_1, m_2 \in M$ such that $\lambda z + m_1 \in E(z^*)$ and $\lambda u + m_2 \in E(z^*)$. Clearly, there exists $m \in M$ such that

$$\|w - (\lambda z + m_1) - (\lambda u + m_2) + m\| < \epsilon$$

and we have that $E(z^*) - E(z^*) + M$ is dense in E. However, if $E(z^*) - E(z^*) + M \neq E$, then since $E(z^*) - E(z^*) + M$ has nonempty interior, there is a continuous linear functional y^* on E such that $\Re(y^*(x)) \leq 1$ for all $x \in E(z^*) - E(z^*) + M$. This would contradict the density proved above. $\square$

Suppose now that E and M are as in the statement of Lemma 2.3.2, and let K denote a locally compact Hausdorff space. Let $\mathcal{M}$ be the subspace of $\mathcal{Z} = C_0(K, E)$ defined by

$$\mathcal{M} = \{F \in \mathcal{Z} : F(K_0) \subset M\}$$

where K_0 is a subset of K. For $t \in K_0$, let ψ_t denote the evaluation function on $\mathcal{Z}$ defined by $\psi_t(F) = F(t)$. We are ready to describe the extreme points of the unit ball of the dual of the quotient space $\mathcal{Z}/\mathcal{M}$.

2.3.3. LEMMA. *(Ströbele) If x^* is an extreme point of $S((E/M)^*)$ and $t_0 \in K_0$, then*

(20)
$$\gamma^* = x^* \circ \psi_{t_0}$$

is an extreme point of $S((\mathcal{Z}/\mathcal{M})^*)$, where $\mathcal{Z} = C_0(K, E)$.

PROOF. Let $\gamma^* = x^* \circ \psi_{t_0}$ where x^* is an extreme point of $B((E/M)^*)$ and $t_0 \in K_0$. By Lemma 2.3.2, $E = E(x^*) - E(x^*) + M$, and in fact it is easily seen that $E = \frac{1}{n}E(x^*) - \frac{1}{n}E(x^*) + M$ for any n. Furthermore, $E(x^*)$ contains elements of arbitrarily large norm. Hence for $F \in \mathcal{Z}$, $F(t_0) \in E$, and there are elements $x_1, y_1 \in \frac{1}{4}E(x^*)$ such that $F(t_0) = x_1 - y_1 + m$ where $m \in M$. We may choose $u \in \frac{1}{4}E(x^*)$ with $\|u\| \geq \|x_1\| + \|F\|$ and if we let $x = x_1 + u$, $y = y_1 + u$, then $x, y \in \frac{1}{2}E(x^*)$ and $F(t_0) = x - y + m$. Furthermore, $\|x\| \geq \|F\|$.

Let U be a neighborhood of t_0 such that $\|F(t) - F(t_0)\| < \frac{1}{2}$ for all $t \in U$. By Urysohn's Lemma, we can find a function $r(t)$ defined on K with values in $[0,1]$ so that $r(t_0) = 1$ and $r(t) = 0$ for $t \in K \backslash U$. If we let $L(t) = r(t)m$, $G(t) = F(t) + r(t)(y - m)$ and $H(t) = r(t)y$, we see that $F = G - H + L$ where $L \in \mathcal{M}$. Since $\gamma^*(H) = x^*(y)$, we see that H belongs to $\mathcal{Z}(\gamma^*)$. We will next show that $G \in \mathcal{Z}(\gamma^*)$.

If $t \in K \backslash U$, then

$$\|G(t)\| = \|F(t)\| \leq \|F\| \leq \|x\|.$$

If $t \in U$, we have

$$\|G(t)\| = \|F(t) - F(t_0) + F(t_0) - r(t)(y - m - x) + r(t)x\|$$
$$\leq \|F(t) - F(t_0)\| + \|(1 - r(t))F(t_0) + r(t)x\|$$
$$< \frac{1}{2} + (1 - r(t))\|F(t_0)\| + r(t)\|x\|$$
$$\leq \frac{1}{2} + \|x\|$$

since $\|F(t_0)\| < \|x\|$. Now $G(t_0) = x$ and since $x \in \frac{1}{2}E(x^*)$, we obtain

$$\|G\| = \Re\gamma^*(g) \le \frac{1}{2} + \|x\| - \Re(x^*(x)) \le 1.$$

Therefore, $G \in \mathcal{Z}(\gamma^*)$, and $F = G - H + L \in \mathcal{Z}(\gamma^*) - \mathcal{Z}(\gamma^*) + \mathcal{M}$.

We apply Lemma 2.3.2 again to see that γ^* is an extreme point of $S((\mathcal{Z}/\mathcal{M})^*) = S(\mathcal{Z}^*) \cap \mathcal{M}^\perp$. □

We now prove a partial converse of Lemma 2.3.3.

2.3.4. LEMMA. *Suppose M is a closed subspace whose complement is 1-complemented in E and $K_0 = K$. Then every extreme point of $B((\mathcal{Z}/\mathcal{M})^*)$ is of the form $z^* \circ \psi_{t_0}$ where $t_0 \in K$ and x^* is an extreme point of $B(E/M)^*)$.*

PROOF. Let $A = \{x^* \circ \psi_t : x^* \in B((E/M)^*), \ t \in K\}$. We first show that A is weak*-closed in $S(\mathcal{Z}/\mathcal{M})^*)$. Assume that $x_\alpha^* \circ \psi_{t_\alpha} \to z^* \in B((\mathcal{Z}/\mathcal{M})^*) = B(\mathcal{Z}^*) \cap \mathcal{M}^\perp$. There are two possibilities for the net $\{t_\alpha\}$. First, suppose $\{t_\alpha\}$ has a cluster point $t_0 \in K$. We may assume (taking subnets if necessary) that $t_\alpha \to t_0$ and $x_\alpha^* \to x^*$ in the weak*-topology of $(E/M)^*$, where $x^* \in B((E/M)^*)$. For any equivalence class $[F]$ in $\mathcal{Z}/\mathcal{M}$,

$$z^*([F]) = z^*(F) = \lim_\alpha (x_\alpha^* \circ \psi_{t_\alpha})(f) = \lim_\alpha x_\alpha^*(f(t_\alpha)) = x^*(f(t_0))$$

so that $z^* = x^* \circ \psi_{t_0} \in A$.

If $\{t_\alpha\}$ has no cluster point in K and $[F] \in \mathcal{Z}/\mathcal{M}$, then for $\epsilon > 0$ there exists a compact set $D \subset K$ such that $\|F(t)\| < \epsilon$ for $t \in K\backslash D$. Since $\{t_\alpha\}$ has no cluster point in K, there is an index α_0 such that $t_\alpha \in K\backslash D$ for $\alpha \ge \alpha_0$. (Otherwise a subnet of $\{t_\alpha\}$ would lie in the compact set D and have a cluster point.) Then

$$|x_\alpha^* \psi_{t_\alpha}(F)| = |x_\alpha^*(F(t_\alpha))| \le \|F(t_\alpha)\| < \epsilon$$

and we conclude that $x_\alpha^* \psi_{t_\alpha} \to 0$ in the w*-topology, and $0 \in A$.

Next we show that the closed convex hull $\overline{co}(A)$ of A must be equal to $B((\mathcal{Z}/\mathcal{M})^*)$. If not, there exist $[F] \in \mathcal{Z}/\mathcal{M}$ and $z_0^* \in S((\mathcal{Z}/\mathcal{M})^*)\backslash\overline{co}(A)$ such that

$$\Re(z_0^*([F])) > sup\{\Re(z^*(F)) : z^* \in \overline{co}(A)\}$$
$$\ge sup\{\Re x^*(F(t)) : x^* \in S((E/M)^*), \ t \in K\}$$
$$\ge \|[F]\|,$$

which is a contradiction. The last inequality requires some justification.

By hypothesis, $E = M \oplus N$ and there exists a projection P onto N along M such that $\|P\| = 1$. Let $G = P \circ F$ so that $F(t) - G(t) = F(t) - PF(t) \in M$ for every t. Hence $F - G \in \mathcal{M}$ and $G \in [F]$. Let $\epsilon > 0$ be given and choose $t \in K$ such that $\|F(t)\| > \|G\| - \epsilon$. We may choose $x^* \in B(E^*)$ so that $|x^*(G(t))| > \|G(t)\| - \epsilon$. Define y^* on E by $y^* = x^* \circ P$. Then $y^* \in M^\perp \cap B(E^*)$ since $\|P\| = 1$. Furthermore,

$$|y^*(F(t))| = |x^*(PF(t))| \ge \|G(t)\| - \epsilon \ge \|G\| - 2\epsilon \ge \|[F]\| - 2\epsilon.$$

Since $y^* \in B((E/M)^*)$, we see that

$$sup\{\Re x^*(F(t)) : x^* \in B((E/M)^*)\} \geq \|[F]\|.$$

By Milman's converse of the Krein-Milman Theorem as quoted by Phelps [251, p.9], the extreme points of $B((\mathcal{Z}/\mathcal{M})^*)$ are contained in A.

The proof is completed by noting that if $z^* = x^* \circ \psi_t$, then z^* is an extreme point only if x^* is an extreme point.   □

We are finally ready to give a general version of the Arens-Kelley Theorem.

2.3.5. THEOREM. (Brosowski and Deutsch) Let E be a (real or complex) normed linear space and $\mathcal{Z} = C_0(K, E)$. Then

$$ext(\mathcal{Z}^*) = \{x^* \circ \psi_t : x^* \in ext(E^*), \, t \in K\}.$$

PROOF. Apply Lemmas 2.3.3 and 2.3.4 with $M = \{0\}$ and $K_0 = K$.   □

We will need one more form of this theorem, one that applies to subspaces of $\mathcal{Z} = C_0(K, E)$.

2.3.6. COROLLARY. If X is a subspace of $C_0(K, E)$, then

$$ext(X^*) \subset \{x^* \circ \psi_t : x^* \in ext(E^*), \, t \in K\}.$$

PROOF. This result follows from the fact that an extreme point of $B(X^*)$ must be the restriction to X of an extreme point of $B(\mathcal{Z}^*)$ and hence of the announced form by Theorem 2.3.5. To see this, suppose y^* is an extreme point of $B(X^*)$. Let $U = \{z^* \in \mathcal{Z}^* : \|z^*\| = 1, \ z^*|X = y^*\}$, and suppose w^* is an extreme point of U. Let $w^* = \frac{1}{2}w_1^* + \frac{1}{2}w_2^*$ where $w_1^*, w_2^* \in B(\mathcal{Z}^*)$. It follows that $\|w_1^*\| = \|w_2^*\| = 1$. For $v_1^* = w_1^*|X$ and $v_2^* = w_2^*|X$, we have $y^* = \frac{1}{2}v_1^* + \frac{1}{2}v_2^*$, and since $v_1^*, \, v_2^* \in B(X^*)$, and y^* is extreme, we conclude that $v_1^* = v_2^* = y^*$. Therefore $w_1^*, w_2^* \in U$ so that $w^* = w_1^* = w_2^*$ and $w^* \in ex(S(\mathcal{Z}^*))$.   □

We are almost ready to state and prove the theorem of Novinger, but before doing so we should say a little about the Choquet boundary. The Choquet Boundary is a subset of Q associated with a subspace M of $C_0(Q)$, and was first introduced by Bishop and DeLeeuw in their work on Choquet's Theorem. In the case were Q is compact and metrizable and M is a function algebra, the Choquet Boundary consists of the peak points for M. Thus it is probably not surprising that the notion shows up in a discussion of isometries.

2.3.7. DEFINITION. If Q is a locally compact Hausdorff space and M is a linear subspace of $C_0(Q)$, the Choquet boundary for M (denoted by $ch(M)$) is the set of all t in Q such that ψ_t is an extreme point of $B(M^*)$.

2.3.8. THEOREM. (Phelps) The Choquet boundary $ch(M)$ is a boundary for M; that is, given $f \in M$, there exists $t \in ch(M)$ such that $|f(t)| = \|f\|$.

PROOF. Given $f \in M$, let t_0 be such that $\|f\| = |f(t_0)|$ and let $A = \{z^* \in B(M^*) : z^*(f) = f(t_0)\}$. Then A is nonempty (since $\psi_{t_0} \in A$), weak*-closed, and convex, so A must have an extreme point y^*. It is straightforward to show that y^* is in fact an extreme point of $B(M^*)$, and by Corollary 2.3.6, $y^* = \lambda\psi_t$ for some $t \in ch(M)$ and scalar λ with $|\lambda| = 1$. Hence $|f(t)| = |\lambda\psi_t(f)| = |f(t_0)| = \|f\|$.   □

2.3.9. DEFINITION. *Let Q be a locally compact Hausdorff space. An element $s_0 \in Q$ is said to be a strong boundary point of a subspace M of $C_0(Q)$ if for each neighborhood U of s_0, and each $\epsilon > 0$, there exists $f \in M$ such that $1 = f(s) = \|f\|$, and $|f(s)| < \epsilon$ for all $s \in Q \backslash U$. The set $\sigma(M)$ of strong boundary points for M is called the strong boundary of M. A subspace M of $C_0(Q)$ is said to be extremely regular if $\sigma(M) = Q$, and M is extremely C-regular if $ch(M) \subset \sigma(M)$.*

Note that an extremely C-regular subspace M must separate the points of $ch(M)$.

2.3.10. THEOREM. *(Novinger)*

(i) *Let Q, K be locally compact Hausdorff spaces and suppose M is a closed subspace of $C_0(Q)$ which separates the points of its Choquet Boundary. If T is a linear isometry of M onto a subspace N of $C_0(K)$, then there is a function h from $ch(N)$ into the unit circle and a function φ from $ch(N)$ onto $ch(M)$ such that*

(21) $$Tf(t) = h(t)f(\varphi(t)) \quad \text{for all } f \in M \text{ and } t \in ch(N).$$

Furthermore, the functions h, φ are continuous at each $t \in ch(N)$ for which $\varphi(t) \in \sigma(M)$. In particular, if M is extremely C-regular, then h and φ are continuous on $ch(N)$.

(ii) *If Q is compact and M is any subspace of $C(Q)$ which separates points of Q and contains the constant functions, then the conclusion of the previous part holds. In this case, h is defined and continuous on all of K and there is a continuous function Φ from K into $S(M^*)$ such that*

(22) $$\overline{h(t)}Tf(t) = \Phi(t)(f) \quad \text{for all } t \in K.$$

PROOF. (i) If $t \in ch(N)$, then $\psi_t \in ext(N^*)$ by Definition 2.3.7 and since T^* is an isometry from N^* onto M^*, it must map extreme points to extreme points. Hence $T^*\psi_t$ is an extreme point of $B(M^*)$ so $T^*\psi_t = \lambda\psi_s$ where $|\lambda| = 1$ and $s \in Q$. It follows that $\overline{\lambda}T^*\psi_t = \psi_s$ is also an extreme point of $B(M^*)$ so $s \in ch(M)$. Let $h(t) = \lambda$ and $\varphi(t) = s$. The function φ is well defined since M separates the points of $ch(M)$. If $f \in M$, then

$$Tf(t) = T^*\psi_t(f) = h(t)\psi_s(f) = h(t)f(s) = h(t)f(\varphi(t))$$

and (21) is satisfied.

If $s \in ch(M)$, then $\psi_s \in ext(M^*)$ and since T^* maps $ext(N^*)$ onto $ext(M^*)$, there must exist some $y^* \in ext(N^*)$ such that $T^*y^* = \psi_s$. Now

$y^* = \lambda\psi_t$ for some $|\lambda| = 1$ and $t \in K$. Again we see that $\overline{\lambda}y^* = \psi_t \in ext(N^*)$ and $t \in ch(N)$. Furthermore, $\psi_s = \lambda T^*\psi_t$ so that $\varphi(t) = s$ and φ maps $ch(N)$ onto $ch(M)$.

To complete the proof we must prove the statement about the continuity of h and φ. First suppose φ is not continuous at $t_0 \in ch(N)$, where $\varphi(t_0) \in \sigma(M)$. Then there is a neighborhood W of $s_0 = \varphi(t_0)$ such that for every neighborhood U of t_0 there is some $t \in U \cap ch(N)$ such that $\varphi(t) \notin W$. Since $s_0 \in \sigma(M)$, given $\epsilon > 0$ there is an $f \in M$ with $f(s_0) = 1 = \|f\|$ and $|f(s)| < \epsilon$ for $s \in ch(M)\backslash W$. Since Tf is continuous at t_0, there exists a neighborhood U_0 of t_0 such that if $t \in U_0 \cap ch(N)$, then

$$h(t)f(\varphi(t)) - h(t_0)f(\varphi(t_0)) < \epsilon.$$

However, since there is a $t \in U_0 \cap ch(N)$ such that $\varphi(t) \in ch(M)\backslash W$, we have

$$\epsilon > |h(t)f(\varphi(t)) - h(t_0)f(\varphi(t_0))| \geq |1 - |f(\varphi(t))|| \geq 1 - \epsilon$$

which is a contradiction for small ϵ.

Next suppose h is not continuous at $t_0 \in ch(N)$ where $\varphi(t_0) \in \sigma(M)$. Then there exists $\epsilon > 0$ such that for every neighborhood U of t_0 there is some $t \in U \cap ch(N)$ with $|h(t) - h(t_0)| \geq \epsilon$. Since $\varphi(t_0) \in \sigma(M)$, there exists an $f \in M$ such that $f(\varphi(t_0)) = 1$. If $r(t) = f(\varphi(t)) - f(\varphi(t_0))$, then continuity of f and φ guarantees a neighborhood U_0 of t_0 such that

$$|r(t)| = |f(\varphi(t)) - f(\varphi(t_0))| < \frac{\epsilon}{4}$$

for $t \in U_0 \cap ch(N)$. Furthermore, there exists a neighborhood U_1 of t_0 such that $|Tf(t) - Tf(t_0)| < \frac{\epsilon}{2}$ for $t \in U_1$. Now let $t \in U_1 \cap U_0 \cap ch(N)$ be such that $|h(t) - h(t_0)| > \epsilon$. We must have

$$\frac{\epsilon}{2} > |Tf(t) - Tf(t_0)| = |h(t)f(\varphi(t)) - h(t_0)f(\varphi(t_0))|$$
$$= |h(t)[f(\varphi(t_0)) + r(t)] - h(t_0)f(\varphi(t_0))|$$
$$= |[h(t) - h(t_0)]f(\varphi(t_0)) + h(t)r(t)|$$
$$\geq ||h(t) - h(t_0)| - |r(t)|| \geq \epsilon - \frac{\epsilon}{4} = \frac{3\epsilon}{4}$$

which is a contradiction.

Clearly, the above arguments hold for all $t \in ch(N)$ when M is extremely C-regular.

(ii) If Q is compact, the map $s \to \psi_s$ is a homeomorphism of $ch(M)$ onto $extS(M^*)$ with its w^*-topology. The function $h(t)$ in the argument of part (i) is just $T1(t)$ and so is continuous since $T1 \in C_0(K)$. The function $\Phi(t) = \overline{h(t)}T^*\psi_t$ is continuous from K into $S(M^*)$ and for $t \in chN$, we have that $\Phi(t) = \psi_s$ for $s \in ch(M)$. In this case, $\Phi(t)(f) = f(\varphi(t))$ from which (21) follows.

$\square$

2.3.11. COROLLARY. *In the notation of Theorem 2.3.10, suppose* M, N *are completely C-regular subspaces of* $C_0(Q)$ *and* $C_0(K)$, *respectively, and suppose* T *is a linear isometry from* M *onto* N. *Then (21) holds where* h *is continuous on* $ch(N)$ *into the unit circle of* $\mathbb{C}$ *and* φ *is a homeomorphism from* $ch(N)$ *onto* $ch(M)$.

PROOF. From Theorem 2.3.10(i) we get h as described in the statement of the corollary and the function φ which is continuous from $ch(N)$ onto $ch(M)$. Since N separates the points of $ch(N)$, φ is one-to-one. We may apply the theorem to T^{-1} to get a modulus one function q continuous on $ch(M)$ and a function β which is continuous from $ch(M)$ onto $ch(N)$. Further, we have from (21) applied to T^{-1} that

$$f(s) = T^{-1}(Tf)(s) = q(s)Tf(\beta(s))$$

for all $s \in ch(M)$ and $f \in M$. However, using (21) applied to T we obtain

$$f(s) = q(s)Tf(\beta(s)) = q(s)h(\beta(s))f((\varphi \circ \beta)(s))$$

so that $|f(s)| = |f((\varphi \circ \beta(s))|$ for all $s \in ch(M)$ and $f \in M$.

Since M is extremely C-regular, we must have $s = \varphi(\beta(s))$ from which we conclude that $\beta = \varphi^{-1}$ and $ch(M)$ is homeomorphic to $ch(N)$. □

2.3.12. COROLLARY. *If* T *is a linear isometry from* $C_0(Q)$ *onto* $C_0(K)$, *then* Q *and* K *are homeomorphic. Furthermore* $Tf(t) = h(t)f(\varphi(t))$ *for all* $t \in K$ *where* h *is continuous on* K *such that* $|h(t)| = 1$ *for all* $t \in K$ *and* φ *is a homeomorphism of* K *onto* Q.

PROOF. In this case, the Choquet boundaries of M and N are Q and K, respectively, and they are homeomorphic by Corollary 2.3.11. □

This corollary is the form of the Banach-Stone Theorem that is sometimes given. The classical case, where Q, K are compact is, of course, included.

2.3.13. COROLLARY. *If* M, N *are extremely C-regular,* Q *is compact, and* $ch(M)$ *is closed, then* $ch(N)$ *is compact. In particular, if* $N = C_0(K)$, *then* $ch(N) = K$ *so* K *is compact.*

2.3.14. COROLLARY. *(Novinger) If* M *separates points of the compact space* Q *and contains the constant functions and if* $N = C_0(K)$, *then* K *is compact and* φ *is a homeomorphism of* K *onto* $ch(M)$.

PROOF. If $N = C_0(K)$ then $K = ch(N)$ and by Theorem 2.3.10 (ii), we have $Tf(t) = h(t)f(\varphi(t))$ for all $t \in K$. Since $1 \in M$, we must have $T1(t) = h(t) \in C_0(K)$. Also, since $|h|$ is constant on K, K must be compact and the continuous mapping φ from the compact set K onto $ch(M)$ is one-to-one and is a homeomorphism. □

2.3.15. COROLLARY. *(Novinger) Suppose* M *separates the points of the compact space* Q *and contains the constant functions. If* $ch(M)$ *is closed in*

Q, then $ch(N)$ is closed in K. If N also separates the points of K and K is compact, then $ch(M)$ and $ch(N)$ are homemomorphic.

PROOF. Under the given conditions, the set $\{\psi_s : s \in ch(M)\}$ is a (weak*-) closed subset of $S(M^*)$ and so its inverse image under the continuous map Φ introduced in (22) must be closed in K. This closed set is just $\varphi^{-1}(ch(M)) = ch(N)$. If K is compact and N separates points of K, then φ is a homeomorphism since it maps the compact set $ch(N)$ onto $ch(M)$. $\square$

When $M = C(Q)$ for Q compact, then $ch(M) = Q$ so that $ch(N)$ is closed by Corollary 2.3.15 and we get the theorem of Holsztynski. In Holsztynski's Theorem, the function φ is defined on the set

$$(23) \qquad D = \bigcup_{s \in Q} \{t \in K : T(Q(s)) \subset K(t)\}$$

where $Q(s)$ and $K(t)$ are as defined in Stone's original proof given in Section 1. Under these conditions, the set D is the same as the Choquet Boundary of N.

There is one more corollary available to us now that treats the situation where M and N are algebras, and it will be of use to us in Chapter 4.

2.3.16. COROLLARY. (deLeeuw, Rudin, Wermer) If M and N are closed subalgebras of $C(Q), C(K)$, respectively, each containing 1, M separates the points of the compact space Q, and T is a linear isometry from M onto N, then

$$(24) \qquad Tf(t) = h(t)T_1f(t) \text{ for all } t \in K,$$

where h is unimodular, and T_1 is an algebra isomorphism of M onto N, with the property that if $\overline{f} \in M$, then $\overline{Tf} = T\overline{f}$.

PROOF. By Theorem 2.3.10, we have a unimodular function h as before and a function φ defined on $ch(N)$ so that (21) holds. In this case, $h = T1$ is defined on all of K. Furthermore, $\overline{h} \in N$. To see that, note that

$$h(t)T(fg)(t) = (Tf)(t)(Tg(t)$$

for all $t \in Ch(N)$. In particular, since T is surjective, there exists $p \in M$ such that $Tp = 1$. Putting this in the equation above, and using the fact that $ch(N)$ is a boundary for N, we get that $hT(p^2) = 1$ so that $\overline{h} = T(p^2) \in N$. Let us define T_1 by

$$T_1f = \overline{h}Tf \text{ for all } f \in M.$$

so that (24) holds. Let $f, g \in M$ and $t \in ch(N)$. It is straightforward to show, using (21), that

$$T_1(fg)(t) = T_1f(t)T_1g(t)$$

for $t \in ch(N)$, and since $ch(N)$ is a boundary for N, the same equality must hold for all $t \in K$. Thus T_1 is multiplicative. In a similar way, we can show that $T_1 \overline{f} = \overline{T_1 f}$, whenever $\overline{f} \in M$. Hence T_1 satisfies the properties claimed in the statement of the corollary, (and is an isometry as well), so that the corollary is proved. □

We conclude this section with some examples.

2.3.17. EXAMPLE.

(i) (Singer) Let $Q = \{1,2,3\}$, $K = \{1,2,3\}$ and $M = \{f \in C(Q) : f(3) = 0\}$. Define $T : M \to C(K)$ by $T(s_1, s_2, 0) = (s_1, s_2, \frac{1}{2}(s_1 + s_2))$. Then T is an isometry; $ch(M) = \{1,2\} = ch(N)$ where N is the range of T. Note that $\psi_3 = \frac{1}{2}(\psi_1 + \psi_2)$ in $B(N^*)$ so that ψ_3 is not extreme in $B(N^*)$. Here $h(t) \equiv 1$ and $\varphi(t) = t$ for $t \in ch(N)$. Observe that $3 \notin D$, the set defined by (23), so $D = ch(N)$ in this case. If we let $K = \{1,2,3,4,5\}$ and define T by $T(s_1, s_2, 0) = (s_1, s_2, -s_1, -s_2, \frac{s_1 + s_2}{2})$, we have $ch(N) = \{1,2,3,4\}$ and φ is not one-to-one.

(ii) (McDonald) Let φ_1, φ_2 be continuous functions on a locally compact space K to a compact space Q and let T be defined on $C(Q)$ by $Tf(t) = \frac{1}{2}[f(\varphi_1(t)) + f(\varphi_2(t))]$ for $t \in K$. If $\Gamma = \{t : \varphi_1(t) = \varphi_2(t)\}$ and $\varphi_1(\Gamma) = Q$, then T is an isometry. If N is the range of T, then $\Gamma = ch(N)$. To see this last statement, suppose $t_0 \in ch(N)$. Then ψ_{t_0} is an extreme point of $B(N^*)$ and so $T^* \psi_{t_0} \in ext(C(Q^*))$. Therefore, $T^* \psi_{t_0} = \lambda \psi_s$ for some $s \in Q$ and $|\lambda| = 1$ by Theorem 2.3.5. Since $T1 = 1$, we have $\lambda = 1$ and

$$\psi_s = \frac{1}{2}\psi_{\varphi_1(t_0)} + \frac{1}{2}\psi_{\varphi_2(t_0)}.$$

However, ψ_s is extreme so $\psi_{\varphi_1(t_0)} = \psi_{\varphi_2(t_0)}$ and $t_0 \in \Gamma$.

On the other hand, if $t_0 \in \Gamma$, let $s = \varphi_1(t_0) = \varphi_2(t_0)$. Now ψ_s is an extreme point of $B(C(Q)^*)$ so there exists $z^* \in ext(N^*)$ such that $\psi_s = T^* z^*$. Thus $f(s) = \psi_s(f) = z^*(Tf)$ and also

$$f(s) = \frac{1}{2}f(\varphi_1(t_0)) + \frac{1}{2}f(\varphi_2(t_0)) = Tf(t_0) = \psi_{t_0}(Tf).$$

We conclude that $z^* = \psi_{t_0}$ and $t_0 \in ch(N)$.

(iii) (Jeang and Wong) Let T be defined from $C_0((0,\infty))$ into $C_0((-\infty,\infty))$ by $Tf(t) = h(t)f(\varphi(t))$ where

$$h(t) = \begin{cases} 1, & \text{if } t \geq 2; \\ t - 1, & \text{if } 0 \leq t \leq 2; \\ -1, & \text{if } t < 0; \end{cases}$$

and $\varphi(t) = |t|$. For $N =$ range of T, it can be seen by the same argument as in the previous example that $(-\infty, 0] \cup [2, \infty)$ is contained in $ch(N)$. For $t \in (0,2)$, it is easily seen that $t \notin D$ where, again, D is the set defined by (23) and since $ch(N) \subset D$ holds in general, we conclude that $(0, 2)$ has no points in common with $ch(N)$ which therefore equals $(-\infty, 0] \cup [2, \infty)$.

(iv) Let $M = \{f \in C[0, 1] : f(0) = f(1) = 0\}$. Take φ_1, φ_2 as in (ii) above and let $Tf = \frac{1}{2}((f \circ \varphi_1) + (f \circ \varphi_2))$, where $Q = K = [0, 1]$, and $\Gamma = [0, 1/2]$. In this case $ch(M) = (0, 1)$ and $ch(N) = (0, 1/2)$ where $N = T(M)$.

2.4. A Theorem of Vesentini

It is well known that the extreme points of the unit ball of $C(Q)$ are those functions f such that $|f(s)| = 1$ for all $s \in Q$. For any subspace M of $C(Q)$, let $\Gamma(M)$ denote the set of functions in M such that $|f(s)| = 1$ for all $s \in ch(M)$.

2.4.1. LEMMA. *If $M \subset C(Q)$, then $\Gamma(M) \subset ext(M)$.*

PROOF. If $f \in \Gamma(M)$, suppose $f = \frac{1}{2}(g + h)$ where $g, h \in B(M)$. As we have seen before, we must have $1 = |f(s)| = |g(s)| = |h(s)|$ for all $s \in ch(M)$, and in fact, $g(s) = h(s) = f(s)$ for all such s. Therefore, $g - h$ is zero on $ch(M)$ and by Theorem 2.3.8, $g - h = 0$ on Q.   □

In the previous section, the technique for describing isometries depended on the fact that the adjoint had to map extreme points to extreme points in the dual unit balls. Here we show that the same description of an isometry as a weighted composition operator follows if the operator maps extreme points of $C(Q)$ to certain extreme points of a subspace of $C(K)$.

2.4.2. THEOREM. *(Vesentini) Let T be a bounded linear operator from $C(Q)$ into $C(K)$, with $\|T\| \leq 1$. If $T(\Gamma(C(Q)) \subset \Gamma(N)$ where N is the range of T, then there exist a continuous map φ from $ch(N)$ to Q and a function $h \in \Gamma(N)$ such that*

(25) $$Tf(t) = h(t)f(\varphi(t))$$

for all $f \in C(Q)$ and $t \in ch(N)$.

PROOF. For $t \in ch(N)$, the map $f \to Tf(t)$ defines a bounded linear functional on $C(Q)$, and so by the Riesz Representation Theorem there is a unique, regular Borel measure μ_t on Q such that $Tf(t) = \int f d\mu_t = \mu_t(f)$ where $\mu_t(f)$ is notation defined by the equality. The hypothesis then implies that

(26) $$|\mu_t(f)| = 1 \quad \text{for all } f \in \Gamma(C(Q)) \text{ and } t \in ch(N).$$

The proof will be completed by proving each of the following statements:

(i) Equation (26) holds for all f in the set $\Lambda(Q)$ of all measurable complex-valued functions on Q which have modulus one almost everywhere with respect to $|\mu_t|$;

(ii) For every $t \in ch(N)$, there is a complex constant $h(t)$ with $|h(t)| = 1$, and a point $s \in Q$ such that $\mu_t(f) = h(t)\psi_s(f)$ for all $f \in C(Q)$; and,

(iii) The function φ defined by $\varphi(t) = s$, where s, t are related as in (ii), is continuous on $ch(N)$ and the function $h(t)$ as defined above is in $\Gamma(N)$.

To begin, let us suppose $g \in \Lambda(Q)$. Then $g(s) = \exp(i\lambda(s))$ where $\lambda(s)$ is measurable. By Lusin's Theorem, there exists, for each n, a continuous function λ_n such that

$$|\mu_t|(\{s \in Q : \lambda_n(s) \neq \lambda(s)\}) < \frac{1}{n}$$

and $|\lambda_n(s)| \leq |\lambda(s)|$ a.e. and it follows that $\lambda_n \to \lambda$ in measure. Hence there is a subsequence $\{\lambda_{n_j}\}$ of $\{\lambda_n\}$ which converges to λ a.e. $(|\mu_t|)$. Now let $g_{n_j}(s) = \exp(i\lambda_{n_j}(s))$ so that $g_{n_j} \in \Gamma(C(Q))$ for each j and $g_{n_j} \to g$ a.e. By Lebesgue's Dominated Convergence Theorem, we have $\mu_t(g_{n_j}) \to \mu_t(g)$ and it follows that $|\mu_t(g)| = 1$ since (26) is true for each g_{n_j}. Hence (26) holds for all $g \in \Lambda(Q)$ and (i) is established.

Next we suppose $t \in ch(N)$ and Q_0 is the support of $|\mu_t|$. For any open set V containing Q_0, $|\mu_t(V)| > 0$. If $s_1, s_2 \in Q_0$ with $s_1 \neq s_2$ and if U is a neighborhood of s_1 such that $s_2 \in Q \backslash \overline{U}$, then

(27) $$|\mu_t|(U) > 0 \text{ and } |\mu_t|(Q \backslash U) > 0.$$

Let $f \in \Gamma(C(Q))$ and α, β given real numbers. Define g on Q by

$$g(s) = \begin{cases} e^{i\alpha} f(s), & \text{if } s \in U, \\ e^{i\beta} f(s), & \text{if } s \in Q \backslash U. \end{cases}$$

Then $g \in \Lambda(Q)$ so that $\left| \int g(s) d\mu_t \right| = 1$. Hence,

$$\left| \int_U e^{i\alpha} f d\mu_t \right|^2 + \left| \int_{Q \backslash U} e^{i\beta} f d\mu_t \right|^2 + 2\Re \left(\int_U e^{i\alpha} f d\mu_t \int_{Q \backslash U} e^{i\beta} f d\mu_t \right) = 1.$$

Differentiation of this equation with respect to α and β, and some algebra with complex numbers leads to the equations

$$\int_U e^{i\alpha} f d\mu_t \overline{\int_{Q \backslash U} e^{i\beta} f d\mu_t} - \overline{\int_U e^{i\alpha} f d\mu_t} \int_{Q \backslash U} e^{i\beta} f d\mu_t = 0$$

and

$$\int_U e^{i\alpha} f d\mu_t \overline{\int_{Q \backslash U} e^{i\beta} f d\mu_t} + \overline{\int_U e^{i\alpha} f d\mu_t} \int_{Q \backslash U} e^{i\beta} f d\mu_t = 0.$$

Therefore,

$$\int_U e^{i\alpha} f d\mu_t \overline{\int_{Q \backslash U} e^{i\beta} f d\mu_t} = 0$$

for all $\alpha, \beta \in \mathbb{R}$ and $f \in \Lambda(Q)$. Let $d\mu_t = h(t)d|\mu_t|$ be the polar decomposition of μ with $h \in \Lambda(Q)$. If we choose $\alpha = \beta = 0$ and $f = 1/h$, we get

$$\int_U d|\mu_t| \int_{Q \setminus U} d|\mu_t| = 0$$

which is a contradiction of (27). It follows that the support Q_0 of μ_t consists of a singleton $\{s\}$ and $\mu_t = h(t)\delta_s$ where $|h(t)| = 1$ and δ_s is the measure with mass 1 concentrated at s. Hence for $f \in C(Q)$,

$$\mu_t(f) = \int f d\mu_t = h(t)f(s) = h(t)\psi_s(f),$$

and we have (ii).

Finally, we define $\varphi(t) = s$, where t, s are related as above and we see that (25) holds. Since the identically one function $1 \in C(Q)$, we have $T1(t) = h(t)$ for all $t \in ch(N)$ and so h can be identified with $T1$ which is defined for all $t \in K$ and belongs to $\Gamma(N)$.

It remains to show that φ is continuous on $ch(N)$. If not, then there is some $t_0 \in ch(N)$ and a neighborhood W of $s_0 = \varphi(t_0)$ in Q such that for every neighborhood U of t_0 there is $t \in U \cap ch(N)$ with $\varphi(t) \notin W$. Let W_0 be a neighborhood of s_0 with $\overline{W_0} \subset W$, and define $f \in C(Q)$ by $f(s) = 1$ for $s \in W_0$ and $f(s) = 0$ for $s \in Q \setminus W$. If ϵ is given with $0 < \epsilon < 1$, then continuity of Tf at t_0 guarantees a neighborhood U_0 of t_0 such that

$$|h(t)f(\varphi(t)) - h(t_0)f\varphi(t_0))| < \epsilon$$

for $t \in U_0 \cap ch(N)$. If $t \in U_0 \cap ch(N)$ and $\varphi(t) \notin W$, then $1 = |f(\varphi(t_0))| < \epsilon < 1$ which is a contradiction.   □

It is interesting that we get the canonical form (25) from the assumption that T maps extreme points of the unit ball of $C(Q)$ to certain extreme points of the unit ball of $N = $ range of T. Such a T, of course, need not be an isometry. For example, the linear transformation $Tf = f(t/2)$ from $C[0, 1]$ onto $C[0, 1]$ satisfies the hypotheses of Theorem 2.4.2, but is not injective and so not an isometry. In this case $\varphi(t) = \dfrac{t}{2}$ is not surjective.

2.4.3. THEOREM. *If T is a linear transformation from $C(Q)$ onto $N \subset C(K)$ and $Tf(t) = h(t)f(\varphi(t))$ where $h \in \Gamma(N)$ and φ is continuous from $ch(N)$ to Q, then T is an isometry if and only if $\varphi(ch(N))$ is dense in Q. Furthermore, T is a surjective isometry if and only if φ is a homeomorphism from K onto Q.*

PROOF. Suppose $\varphi(ch(N))$ is dense in Q and $Tf = 0$ for some $f \in C(Q)$. Let $s \in Q$, and suppose $\epsilon > 0$ is given. By continuity of f there is a neighborhood U of s so that $|f(s) - f(s')| < \epsilon$ for $s' \in U$. Since U contains $\varphi(t)$ for some $t \in ch(N)$, we gave

$$\epsilon > |f(s) - f(\varphi(t))| = |f(s) - \overline{h(t)}Tf(t)| = |f(s)|.$$

We conclude that $f(s) = 0$ so that T is one-to-one, and so is an isometry.

On the other hand if $\varphi(ch(N))$ is not dense, we can find a nonzero $f \in C(Q)$ which is zero on the closure of $\varphi(ch(N))$ so that $Tf = 0$ and T is not injective.

If T is a surjective isometry, then $ch(N) = K$ and since $C(K)$ separates, φ is one-to-one, and so a homeomorphism from the compact set K to the compact set Q. If φ is a homeomorphism and $g \in C(K)$, then $f \in C(Q)$ where $f(s) = h(\varphi^{-1}(s))f(\varphi^{-1}(s))$, and $Tf = g$.   □

Note that if T is an isometry from $C(Q)$ onto $C(K)$, then T must map extreme points to extreme points and so satisfies the hypotheses of Theorem 2.4.2. Hence, the classical Banach-Stone Theorem is a corollary of Theorems 2.4.2 and 2.4.3, and we have still another proof of that result.

If in the hypotheses of Theorem 2.4.2 we require that T map $\Gamma(C(Q))$ into $\Gamma(C(K))$, then (25) holds for all $t \in K$. This is what Vesentini actually proved. However, in this case, $ch(N) = K$ even though N is not necessarily all of $C(K)$. We can get N as the range of weighted composition operator on $C(Q_0)$ where $Q_0 = \varphi(K)$.

2.4.4. THEOREM. *Suppose T is a weighted composition operator defined from $C(Q)$ into $C(K)$ by $Tf(t) = h(t)f(\varphi(t))$ for $t \in K$ where h is continuous with $|h(t)| = 1$ for all $t \in K$ and φ is continuous from K onto Q. If N is the range of T, then $ch(N) = K$.*

PROOF. If $t \in K$, then $\psi_{\varphi(t)} \in ext(C(Q)^*)$. Thus $\psi_{\varphi(t)} = T^*z^*$ where $z^* \in ext(N^*)$. This we know since T must be an isometry. It follows that $z^* = \overline{h(t)}\psi_t$ as elements of N^* and so ψ_t is an extreme point of $B(N^*)$. Hence, $t \in ch(N)$.   □

2.5. Notes and Remarks

Banach first proved his version of the Banach-Stone Theorem in the context of a study of isometries and was apparently interested in actual descriptions of these transformations. The development of the result in the next few years was motivated rather by the desire of investigators to characterize the topological space K according to various structures associated with the space $C(K)$. Thus Stone proved his version in the context of "characterizing completely the algebraico-topological structure of the function rings for bicompact H-spaces" [295].

In this spirit, Gelfand and Kolmogoroff [114] showed that as a ring alone $C(K)$ characterizes K. (Šilov [285] did this for the case where K is compact.) Eilenberg [92], Arens and Kelley [16], and Myers [228], [229], [230] considered only the Banach space structure of $C(K)$, and Kaplansky [166] showed that as a lattice alone, $C(K)$ determines K. He showed that all of the earlier results were subsumed by his theorem. Related papers include those of M. and S. Krein [186], Kakutani [159], Stone [296], Clarkson [75], Milgram [223], and Fan [97].

One type of generalization that has been popular is to weaken the assumptions about the topological spaces. Thus Jarosz and Pathak consider isometries on subspaces of $C(Q)$ given other norms [150] while Araujo and Jarosz [12] obtain the canonical description of isometries on metric spaces of unbounded continuous functions defined on noncompact topological spaces. Similarly, Bachir [19] has extended the Banach-Stone theorem to certain subspaces of the bounded continuous functions on a complete metric space.

In another direction, Hyers and Ulam [140] showed that if there exists some ϵ-isometry from $C(Q, \mathbb{R})$ to $C(K, \mathbb{R})$ then Q and K are homeomorphic. Amir [5] and Cambern [51], [52] showed that the presence of small bound isomorphisms from $C(Q)$ to $C(K)$ imply that Q and K are homeomorphic. Papers related to this approach include Bourgin [45], [46], Cengiz [63], Benyamini [25] , and Jarosz [145]. A brief survey of such results can be found in [102].

There are good discussions of the Banach-Stone Theorem in the books of Semadeni [282, Sec.7,Sec.22], Behrends [23], and Jarosz and Pathak [151].

The various proofs of the Banach-Stone Theorem usually involve the invariance under isometries of certain objects associated with $C(Q)$. Behrends [23] sketches the types of arguments involved when the objects are extreme points, T-sets, and M-ideals, respectively. A paper by Cutland and Zimmer [80] gives a proof using nonstandard analysis.

Finally, we remark here that the early papers on the Banach-Stone Theorem consider only the case where the continuous functions are real valued. The proof given by Dunford and Schwartz [89] treats the complex case.

Eilenberg's Theorem. The notion of the star of an element of a Banach space was first given by Eilenberg [92], who used it to identify peak points. Thus Lemma 2.2.2 and Theorems 2.2.3, 2.2.4 are given by him in the 1942 paper. Eilenberg did not give the details involved in actually establishing the canonical form (18), leaving that to follow in the same way as Stone had given it. But Stone [295] showed that $Vf \geq f$ when $f \geq 0$ (here we assume $V1 = 1$), and his argument does not carry over to the complex case. We obtain the form by showing that $Vg(\psi(s)) = 0$ if $g(s) = 0$. This was shown by Holsztynski [133] but with a different argument.

Eilenberg's paper inspired the work of Myers [228], [229], [230], who introduced the definition of a T-set. A T-set is a subset of a Banach space X which is maximal with respect to the property that if $x_1, x_2, \ldots x_n$ are in the set then $\| \sum x_j \| = \sum \|x_j\|$. This property shows up in Stone's proof, and the intersection of a T-set with the surface of the unit ball is a maximal convex set as used by Eilenberg [92]. If the star of an element x is convex, then it is contained in a T-set, namely the half-cone generated by $st(x)$. The T-sets are important in Jerison's extension [153] of the Banach-Stone Theorem to the vector valued case, which will be treated in Chapter 7 of the second volume.

The Nonsurjective Case. This section makes use of what we call the "extreme point" method of finding isometries. This method takes advantage of the fact that isometries must map extreme points to extreme points, where usually it is the extreme points of the dual unit balls that are involved. Our approach to the characterization of extreme points in a very general setting follows Strobele [297]. We have cast the statement (Lemma 2.3.3) in a form applying to certain quotient spaces, but it does not differ in any essential way from the statement of Theorem 2 in [297]. This approach requires the Lemmas 2.3.1 and 2.3.2 which are due to Buck [48], [49] and Phelps [250].

Strobele [297] asked about a proof of the converse of his Theorem 2 and our Lemma 2.3.4 does that under special assumptions. Theorem 2.3.5 we have attributed to Brosowski and Deutsch [47] since they seem to be the first to state and prove the result in this form in its entirety. Strobele proved that every functional of the form $x^* \circ \psi_t$ is an extreme point of $B(C_0(K, E)^*)$ by using Lemma 2.3.3 as we have indicated. Versions of the Arens-Kelley Theorem have been obtained by a variety of authors. Arens and Kelley [16] were probably the first to get the result for $C(K)$ where K is compact. The proof given by Dunford and Schwartz [89, p.441] is well known and avoids measure theoretic arguments. Singer [287] gave a proof for $C(K, E)$ where K is compact and E a Banach space. Lau [194] quotes the result in this case and attributes it to Lazar [195]. Brosowski and Deutsch [47] give credit to P. D. Morris for helping in their proof. Behrends [23] gives a proof in the context of Banach modules; Tate [302] gives a characterization for certain commutative algebras over $\mathbb{R}$.

The proof of Corollary 2.3.6 follows the argument given by Hoffman [132, p.145]. The fact that an extreme point of $B(X^*)$ is a restriction to X of an extreme functional on $C_0(K, E)$ is due originally to Singer [286] and the example of 2.3.17(i) was given by Singer to show that an extension of a functional might be an extreme point even though the original is not. The statement of the Corollary is also given in [47].

The Choquet Boundary was first mentioned by that name in a paper of Bishop and de Leeuw [36] on Choquet's Theorem. They state a lemma which they say was announced by Bauer [22]. Some of the ideas were no doubt present in the work of Krein and Milman [187]. The fact that the Choquet Boundary consists of peak points for M in the compact, metrizable case is due to Bishop [35]. (See also [251].) A good discussion of the ideas and history are given by Semadeni [282, pp.412-413,428]. The definition we have used, stated in Definition 2.3.7, was first given by Novinger [237] in this general form. The proof of Theorem 2.3.8 follows the argument of Theorem 6.3 in [251]. The latter reference is, of course, an excellent one for information about the Choquet boundary. Theorem 2.3.10 (ii) is the theorem stated and proved by Novinger [237]. Theorem 2.3.10 (i) is our slight perturbation of Novinger's theorem. Although the statement is a bit ugly, it does seem to indicate what is really necessary for the proof to work. The strong boundary was studied by Araujo and Font [10],[11], who showed, for example, that the strong boundary for a point-separating closed subalgebra A of $C_0(Q)$ is dense in the Šilov boundary of A. Recall that the Šilov boundary of a space A is the smallest closed boundary for A. It is also known that the Šilov boundary of such a space is all of Q [110]. The notion of *extremely regular* subspaces was given by Cengiz [63] and our alteration of his definition to extremely C-regular involves only the replacement of Q by $ch(M)$. Cengiz showed that extremely regular subspaces of $C_0(Q)$ arise, for example, as kernels of nonzero, continuous complex-valued finite regular Borel measures on Q, and that $C_0(Q)$ has proper extremely regular subspaces whenever Q is not *dispersed*. Note that Theorem 2.3.10 (i) applies to the examples (i), (iii), (iv) of 2.3.17.

The case where $M = C(Q)$ (in either 2.3.10 (i) or (ii)) gives the theorem of Holsztynski [133] who was the first to obtain a form of the Banach-Stone Theorem in the case where the isometry T is not surjective. Geba and Semadeni [113] had obtained the same result under the assumption that T is *isotonic*; that is, $Tf \geq 0$ if $f \geq 0$. Holszynski's proof was quite different than the one we have given. Pelczynski [249] gave a proof of Holsztynski's theorem and applied it to his study of complemented subspaces of $C(Q)$.

As early as 1948, Myers [228] had considered isometries from a subspace of $C(Q)$ into $C(K)$. A subspace M of $C(Q)$ was defined to be *completely regular* over Q if for every closed subset D of Q and every $s_0 \in Q \backslash D$, there exists $f \in M$ such that $f(s_0) = \|f\|$ and $\sup_{s \in D} |f(s)| < \|f\|$. He then proved that if a completely regular subspace M of $C(Q)$ is isometrically isomorphic to a completely regular subspace N of $C(K)$, where Q and K are compact, then Q and K are homeomorphic. He gave conditions sufficient that there exists Q such that a given Banach space be equivalent to a closed, completely regular subspace of $C(Q)$, and also gave necessary and sufficient conditions that a given Banach space be equivalent to $C(Q)$ for a compact Q. Myers was not concerned with giving an explicit description of a given isometry, but his papers [228], [229], [230] contain many ideas that are useful in study of the Banach-Stone Theorem.

Corollaries 2.3.14 and 2.3.15 may be found in the paper of Novinger [237]. Corollary 2.3.16 was proved by de Leeuw, Rudin, and Wermer for the case where $M = N$ [85]. (See also, Hoffman [132, pp.144-145].) Novinger [237] gives this corollary under the assumption that $T1$ is identically one, so that T itself is multiplicative. In fact, Nagasawa [231] was probably the first to show that function algebras are isometric if and only if they are isomorphic as algebras. Kulkarni and Limaye [189, Chap.5] studied isometries between real function algebras.

The set D given by (23) can be shown to be equal to the Choquet Boundary of N in the case $M = C(Q)$ with Q compact. The proof requires the fact that $f(s) = 0$ implies that $Tf(t) = 0$ which was proved in this setting by Holsztynski [133]. (See also [282].) It is easy to show that $ch(N)$ is contained in D in the setting of Theorem 2.3.10.

McDonald [220] gave a special case of the class of examples given in 2.3.17(ii). His purpose was to show that not every isometry from $C(Q)$ into $C(K)$ need be a weighted composition operator. In his paper he is concerned with linear isometries from $C(Q)$ into $C(K)$ under certain conditions on Q

and K, and when they are weighted composition operators, at least on special subsets of K. Example 2.3.17(iii) is due to Jeang and Wong [**152**], who give it as an example of an isometry from $C_0(Q)$ to $C_0(K)$ that cannot be extended to an isometry from $C(Q_\infty)$ to $C(K_\infty)$ where Q_∞ and K_∞ are the one point compactifications of Q and K, respectively. In that paper Jeong and Wong prove Theorem 2.3.10(i) in the special case where $M = C_0(Q)$. They also show that every disjointness preserving operator is a weighted composition operator when restricted to a special subset of K, which is a generalization of a result of Jarosz [**148**]. Holsztynski's form of the theorem has been considered for the vector valued case by Cambern [**54**].

Finally, we want to state a theorem from the paper of Araujo and Font [**10**]. We will need some terminology and notation. A subspace M of $C_0(Q)$ is said to be *separating* (*strongly separating*) if for every pair s, t of distinct points in Q there is a function $f \in M$ such that $f(s) \neq f(t)$ ($|f(s)| \neq |f(t)|$). By $\xi(M)$ we will mean the set of all $s_0 \in Q$ such that for each neighborhood U of s_0, there exists $f \in M$ such that $|f(s)| < \|f\|$ for all $s \in Q \backslash U$. By $\xi_0(M)$ will be meant the set

$$\xi_0(M) = \xi(M) \cap \{s \in Q : f(s) \neq 0 \text{ for some } f \in M\}.$$

2.5.1. THEOREM. *(Araujo and Font) Let T be a linear isometry from a strongly separating subspace M of $C_0(Q)$ onto such a subspace N of $C_0(K)$. Then there exists a homeomorphism φ of $\xi_0(N)$ onto $\xi_0(M)$ and a continuous unimodular map h defined on $\xi_0 N)$ such that*

(28) $$(Tf)(t) = h(t)f(\varphi(t)) \text{ for all } t \in \xi_0(N), \quad f \in M.$$

It is true also that the map φ above maps $ch(N)$ onto $ch(M)$, and so the above theorem is an extension of Corollary 2.3.11.

If for a given subspace M it is known that the Šilov boundary, $\partial(M)$, of M is not empty, there is a nice alternate description of it.

2.5.2. PROPOSITION. *If M is a linear subspace of $C_0(Q)$, and ∂M is not empty, then $\partial M = \xi(M)$.*

PROOF. Let $s_0 \in \partial M$ and let U be a neighborhood of x_0. Since $Q \backslash U$ is closed and does not contain ∂M, it cannot be a boundary. Thus there is some $f \in M$ so that $|f(s)| < \|f\|$ for all $s \in U$. On the other hand, if $s_0 \in \xi(M) \backslash \partial M$, there is a neighborhood U which does not intersect ∂M. The $f \in M$ guaranteed by $s_{(0)} \in \xi(M)$ does not assume its norm on ∂M, contradicting the fact that it is a boundary.   □

In fact, ∂M is nonempty and coincides with the closure of $ch(M)$ when M is strongly separating. If M is strongly separating and has the property that for each $s \in Q$, there is some $f \in M$ with $f(s) \neq 0$, M is sometimes called a strongly separating *function* subspace. In this case, $\xi_0(M) = \partial M$. Hence, by combining 2.5.2 and 2.5.1 we obtain

2.5.3. THEOREM. *In Theorem 2.5.1, if M, N are strongly separating function subspaces, then the map φ in (28) is a homeomorphism from ∂N onto ∂M.*

This is very nice result and will be of use in Chapter 6.

Everything that we have been discussing in the last few paragraphs is in the paper of Araujo and Font [10], although we have changed the notation somewhat. The authors' arguments are not based on the usual concepts of T-sets or extreme points of the dual balls, but are closer to the approach of Stone. The proofs are lengthy and we have chosen to omit them.

Vesentini's Theorem. The characterization of the extreme points of $C(Q, \mathbb{R})$ was given by Arens and Kelley [16] and mentioned also by Kaplansky [166]. Vesentini [310, Lemma 1] gives a proof of the result which holds in the complex case as well. Theorem 2.4.2 is a slight alteration of the theorem proved by Vesentini [310] in that we demand only that T map extreme points of $C(Q)$ into the special extreme points of N as given by Lemma 2.4.1. This results in a slightly weaker conclusion, but it applies to examples such as those of McDonald given in 2.3.17(ii).

Vesentini [310] was interested in the role of surjectivity for isometries of $C(Q)$ and for holomorphic isometries for the Caratheodory-Kobayashi differential metric of the open unit ball of $C(Q)$. He gives an example of a nonlinear isometry of $C(Q)$ into $C(Q)$ which fixes the zero element. (See Section 3 of Chapter 1.)

In [311], Vesentini obtains an analogue to Theorem 2.4.2 where Q and K are locally compact, Hausdorff spaces and the spaces $C(Q), C(K)$ are endowed with the topology of uniform convergence on compact sets. He also investigates locally equicontinuous semigroups of linear operators from $C(Q)$ into $C(K)$.

CHAPTER 3

The L^p Spaces

3.1. Introduction

Having examined the nature of the isometries on continuous function spaces, it is natural to turn to the L^p-spaces. If φ is a homeomorphism of $[0,1]$ onto itself, the simple composition operator $Tf(t) = f(\varphi(t))$ is not necessarily an isometry on $L^p[0,1]$ For example, let the function φ be given by

$$\varphi(t) = \begin{cases} \frac{t}{2}, & \text{if } 0 \le t \le \frac{1}{2} \\ \frac{3}{2}t - \frac{1}{2}, & \text{if } \frac{1}{2} < t \le 1. \end{cases}$$

The problem, of course, is that the homeomorphism here is not measure preserving. To get an isometry we must multiply the composition operator by the function h where $h(t) = (\frac{3}{2})^{\frac{1}{p}}$ for $\frac{1}{2} < t \le 1$ and $(\frac{1}{2})^{\frac{1}{p}}$ for $0 \le t \le \frac{1}{2}$. Thus $Tf(t) = h(t)f(\varphi(t))$ is an isometry, and as we shall see, all the isometries on L^p take this form. Banach, not surprisingly, was the first to describe the isometries on $L^p[0,1]$, although he did not give the full proof for this case. This was provided by Lamperti who characterized the isometries on L^p of a general and σ-finite measure space, where $0 < p < \infty$, $p \ne 2$. In this chapter we give a complete exposition of Lamperti's proof, as well as Hardin's results on isometries of subspaces of L^p and how they can be extended to larger subspaces. This work is built on a generalization of Rudin's theorem on L^p isometries and equimeasurability and we will discuss that theorem as well.

Banach proved that an isometry on L^p has the property that it maps functions with disjoint support onto functions with disjoint support. This property lies at the base of the proofs to be given in this chapter. It actually arises in a variety of situations, and usually leads to a description of the operator as a weighted composition operator. Let us describe how it works in a somewhat general setting.

Let (Ω, Σ, μ) be a finite measure space and suppose E is a Banach space of measurable functions defined on Ω which contains the characteristic functions of measurable sets and in which the simple functions are dense. Suppose U is a linear transformation on E such that $f \cdot g = 0$ a.e. implies $Uf \cdot Ug = 0$ a.e. Let $A\epsilon\Sigma$ and $A^c = \Omega \backslash A$. Then $1 = \chi_A + \chi_{A^c}$ and $U1 = U(\chi_A) + U(\chi_{A^c})$. Since $U(\chi_A)$ and $U(\chi_{A^c})$ have disjoint supports, it follows that

$$U(\chi_A) = h \cdot \chi_{T(A)}$$

where $h = U1$ and $T(A) = supp U(\chi_A)$. The set function T can be shown to be a "regular set isomorphism" on Σ and it gives rise to a linear transformation denoted by T_1 and called the transformation induced by T, where $T_1(\chi_A) = \chi_{T(A)}$ and $U_{\chi_A} = h T_1(\chi_A)$. Extension to simple functions is clear and by limits (because of the density) we get the form

$$(29) \qquad\qquad U f = h T_1(f).$$

Under the right conditions, the set isomorphism T can be given by a point mapping φ and the operator then takes the familiar form,

$$(30) \qquad\qquad U f(t) = h(t) f(\varphi(t)).$$

Finally, in this chapter, we consider the case where $p = 2$ by looking at Bochner's characterization (a generalization of Plancherel's Theorem) of unitary operators in terms of integrable kernels, and then Merlo's further generalization to isometries from L^p to L^q utilizing what he calls Bochner kernels. As in Chapter 2, we consider here only spaces of scalar valued functions, leaving the vector-valued case to a later chapter.

3.2. Lamperti's Results

We begin by establishing a general form of Clarkson's inequality. Recall that a real function ψ is said to be convex if $\psi(\frac{1}{2}(t+s)) \leq \frac{1}{2}(\psi(t) + \psi(s))$.

3.2.1. LEMMA. *Let φ be a continuous, strictly increasing function defined on the nonnegative reals, with $\varphi(0) = 0$. Assume that $\varphi(\sqrt{t})$ defines a convex function. If z and w are complex numbers, then*

$$(31) \qquad \varphi(|z+w|) + \varphi(|z-w|) \geq 2\varphi(|z|) + z\varphi(|w|).$$

If $\varphi(\sqrt{t})$ is concave, the reverse inequality holds and in the case that convexity or concavity is strict, equality holds in (31) if and only if $zw = 0$.

PROOF. By the convexity of $\varphi(\sqrt{t})$, we have

$$\varphi(\frac{|z+w|^2 + |z-w|^2}{2}) \leq \frac{1}{2}\varphi(\sqrt{|z+w|^2}) + \varphi(\sqrt{|z-w|^2})$$
$$= \frac{1}{2}(\varphi(|z+w|) + \varphi|z-w|)).$$

Since φ^{-1} is increasing,

$$(|z^2| + |w^2|)^{\frac{1}{2}} = (\frac{|z+w|^2 + |z-w|^2}{2})^{\frac{1}{2}} = \varphi^{-1}[\varphi((\frac{|z+w|^2 + |z-w|^2}{2})^{\frac{1}{2}})]$$
$$(32) \qquad \leq \varphi^{-1}(\frac{1}{2}(\varphi(|z+w|) + \varphi(|z-w|))$$

For any continuous convex function ψ, which is zero at zero, if $r < s < t$, then

$$\frac{\psi(s) - \psi(r)}{s - r} \leq \frac{\psi(t) - \psi(r)}{t - r}$$

It follows from the convexity of $\varphi(\sqrt{t})$ that $\frac{t^2}{\varphi(t)}$ is decreasing, and in fact strictly decreasing if the convexity is strict. Now we observe that for any function f defined on the positive reals, if $t^{-1}f(t)$ is decreasing, we have

$$(t+s)^{-1}f(t+s) \le t^{-1}f(t)$$

and

$$(t+s)^{-1}f(t+s) \le s^{-1}f(s).$$

Hence,

$$(t+s)[(t+s)^{-1}f(t+s)] \le t \cdot t^{-1}f(t) + s \cdot s^{-1}f(s)$$

or simply

$$f(t+s) \le f(t) + f(s).$$

We apply this fact to the function $f = [\varphi^{-1}]^2$ which leads to

$$(\varphi^{-1}[\varphi(|z|) + \varphi(|w|)])^2 \le [\varphi^{-1}\varphi(|z|)]^2 + [\varphi^{-1}(|w|)]^2.$$

Upon taking square roots we arrive at the inequality

(33) $$\varphi^{-1}[\varphi(|z|) + \varphi(|w|)] \le (|z|^2 + |w|^2)^{\frac{1}{2}}.$$

Combining (32), (33) and the fact that φ^{-1} is increasing we obtain (31). If the convexity of $\varphi(\sqrt{t})$ is strict, the inequalities we have obtained are also strict, unless z or w is zero. For the case where $\varphi(\sqrt{t})$ is concave, the inequalities are reversed. $\qquad\square$

Given a σ-finite measure space (Ω, Σ, μ) and a function φ, we can define a functional I on the measurable functions by

(34) $$I(f) = \int \varphi(|f|) \, d\mu.$$

Let L^φ denote the measurable functions f for whch $I(f)$ is finite.

3.2.2. THEOREM. *(Lamperti, Clarkson) Let φ be a continuous, strictly increasing function defined on $[0, \infty)$, with $\varphi(0) = 0$, and $\varphi(\sqrt{t})$ convex. If $f + g$ and $f - g$ are in L^φ, then*

(35) $$I[f+g] + I[f-g] \ge 2I[f] + 2I[g].$$

If $\varphi(\sqrt{t})$ is concave, the reverse inequality is true. If the convexity or concavity of $\varphi(\sqrt{t})$ is strict, then equality holds in (35) if, and only if, $f(t)g(t) = 0$ almost everywhere.

PROOF. Let φ be given as in the hypotheses and suppose $\varphi(\sqrt{t})$ is convex. If $f + g$ and $f - g$ are in L^φ, then

$$\int \{\varphi(|f(t) + g(t)|) + \varphi(|f(t) - g(t)|) - 2\varphi(|f(t)|) - 2\varphi(|g(t)|)\} d\mu \ge 0$$

since the integrand is nonnegative by 3.2.1. This inequality is clearly equivalent to (35). For equality to occur, we must have the integrand equal to zero almost everywhere and as we have seen, this happens if and only if $f(t)g(t) = 0$ almost everywhere. The case where $\varphi(\sqrt{t})$ is concave is treated similarly. $\square$

We intend now to make use of Theorem 3.2.2 in the case for which $\varphi(t) = t^p$, where $0 < p < \infty$. In case $p > 2$, $t^{\frac{p}{2}}$ is convex while it is concave for $0 < p < 2$. The key will be to use the case where equality occurs in (35), which then leads to the disjoint support property.

3.2.3. DEFINITION. *Let* $(\Omega_1, \Sigma_1, \mu_1)$, $(\Omega_2, \Sigma_2, \mu_2)$ *denote measure spaces. A set map T from Σ_1 into Σ_2 defined modulo null sets is called a* regular set isomorphism *if*

(i) $T(\Omega_1 \backslash A) = T\Omega_1 \backslash TA$ *for all* $A \in \Sigma_1$.
(ii) $T(\bigcup_1^\infty A_n) = \bigcup_{n=1}^\infty TA_n$ *for disjoint* $A_n \in \Sigma_1$.
(iii) $\mu_2(TA) = 0$ *if and only if* $\mu_1(A) = 0$.

We say T is *measure preserving* if $\mu_1(A) = \mu_2(TA)$ for all $A \in \Sigma_1$. All set equalities and containments are understood to be modulo null sets. It may be useful to observe some properties of a regular set isomorphism T.

3.2.4. REMARKS. *Properties of a regular set isomorphism.*

(i) *If* $A \subseteq B$, *then* $TA \subseteq TB$,
(ii) *The item (ii) above holds even for nondisjoint* A_n,
(iii) $T(\bigcap_1^\infty A_n) = \bigcap_1^\infty (TA_n)$ *for all sequences* $\{A_n\}$ *in* Σ_1,
(iv) $TA \cap TB = \emptyset$ *if and only if* $A \cap B = \emptyset$.
(v) *The regular set isomorphism T induces a unique linear transformation T_1 from the $\Sigma_1 -$ measurable functions into the $\Sigma_2 -$ measurable functions (actually equivalence classes) for which the following hold:*
  (a) $T_1 \chi_E = \chi_{T(E)}$ *for all* $E \in \Sigma_1$;
  (b) *if* $\{f_n\}$ *is a sequence of Σ_1-measurable functions such that $f_n \to f$ a.e., then $T_1 f_n \to T_1 f$ a.e.;*
  (c) $(T_1 f)^{-1}(B) = T_1(f^{-1}(B))$ *for every Borel set B in* $\mathbb{F}$;
  (d) $T_1(fg) = (T_1 f)(T_1 g)$ *and* $T_1 \bar{f} = \overline{T_1 f}$ *for all Σ_1-measurable functions f, g.*
  (e) $T_1(|f|) = |T_1(f)|$ *for all Σ_1-measurable functions.*

The operator T_1 discussed in (v) above will be of great importance in what follows. One method of defining T_1 is to let

$$(36) \qquad T_1 f(t) = s, \quad \text{if } t \in (\cap_{r>s} T(f^{-1}(-\infty, r])) \backslash \cup_{r<s} T(f^{-1}(-\infty, r]))$$

where the $r's$ represent rational numbers. The properties (v)a,(v)b,(v)c,(v)d, and (v)e can be verified directly from this definition. Note that T_1 is a positive operator.

Another approach, where we assume μ_1 is finite, is to consider the measure $\mu_1 \circ T^{-1}$ defined on $T(\Sigma_1)$ which is absolutely continuous with respect to μ_2 by part (iii) of Definition 3.2.3. By the Radon-Nikodym Theorem, there exists a $T(\Sigma_1)$-measurable function g defined on $T(\Omega_1)$ such that

$$\mu_1(T^{-1}(B)) = \int_B g(t) d\mu_2(t) \text{ for all } B \in T(\Sigma_1).$$

Let $g(t) = 0$ for all $t \in \Omega_2 \backslash T(\Omega_1)$, so that we have g is $T(\Sigma_1)$-measurable when restricted to $T(\Sigma_1)$ and Σ_2-measurable on Ω_2. Define a measure ν on Σ_2 by

$$\nu(B) = \int_B g(t) \mathrm{d}\mu_2.$$

Then

$$\nu(\Omega_2) = \int_{\Omega_2} g(t) d\mu_2 = \int_{T(\Omega_1)} g(t) d\mu_2 = \mu_1(\Omega_1) < \infty.$$

Thus ν is a finite measure and if we let $T_1(\chi_A) = \chi_{TA}$ and extend T_1 linearly, we see that it is an isometry on the class of simple functions in $L^p(\Omega_1, \Sigma_1, \mu_1)$ into $L^p(\Omega_2, \Sigma_2, \nu)$. By the density of the simple functions, we can extend T_1 to an isometry on all of $L^p(\Omega_1, \Sigma_1, \mu_1)$. Note that this transformation must be the same as the one defined by (36) since it agrees on the simple functions and 3.2.4(v)b shows it must extend in the same way.

If h is a Σ_2-measurable function such that

$$\int_{TA} |h(t)|^p d\mu_2 = \int_{TA} g(t) d\mu_2$$

for every $A \in \Sigma_1$, where g is as defined above, then for any nonnegative Σ_2-measurable function F,

$$\int_{TA} F(t)|h(t)|^p d\mu_2 = \int_{TA} F(t)g(t) d\mu_2 = \int_{TA} F(t) d\nu.$$

Letting $F = |T_1 f|^p$, we see that

(37) $$Sf(t) = h(t)T_1 f(t)$$

defines an isometry from $L^p(\Omega_1, \Sigma_1, \mu_1)$ into $L^p(\Omega_2, \Sigma_2, \mu_2)$.

We are now ready to state and prove Lamperti's theorem. We assume that the measure spaces are σ-finite.

3.2.5. THEOREM. *Suppose U is a linear isometry from $L^p(\Omega_1, \Sigma_1, \mu_1)$ into $L^p(\Omega_2, \Sigma_2, \mu_2)$ where $1 \le p < \infty$, $p \ne 2$. Then there exists a regular set isomorphism T from Σ_1 into Σ_2 and a function h defined on Ω_2 so that*

(38) $$Uf(t) = h(t)T_1 f(t)$$

where T_1 denotes the transformation induced by T and h satisfies

(39) $$\int_{TA} |h|^p d\mu_2 = \int_{TA} \frac{d(\mu_1 \circ T^{-1})}{d\mu_2} d\mu_2 = \mu_1(A) \text{ for each } A \in \Sigma_1.$$

Conversely, for any h and T as above, the operator U satisfying (38) is an isometry.

PROOF. First we assume that $\mu_1(\Omega_1) < +\infty$. Given $A \in \Sigma_1$, we let $TA = \operatorname{supp}U\chi_A$. If $A, B \in \Sigma_1$ and $A \cap B = \emptyset$, then

$$\| \chi_A + \chi_B \|_p^p + \| \chi_A - \chi_B \|_p^p = 2 \| \chi_A \|_p^p + 2 \| \chi_B \|_p^p.$$

Since U is an isometry, we have

$$\| U\chi_A + U\chi_B \|_p^p + \| U\chi_A - U\chi_B \|_p^p = 2 \| U\chi_A \|_p^p + 2 \| U\chi_B \|_p^p$$

and by Theorem 3.2.2, equality in this "Clarkson Inequality" implies that $U\chi_A$ and $U\chi_B$ have disjoint supports (at least modulo sets of measure zero). It follows that

$$T(A \cup B) = TA \cup TB,$$

where again, as throughout this argument, the set equalities are understood to be correct except for sets of zero measure. From this, it is easily seen that $T(\Omega \backslash A) = T\Omega \backslash T(A)$, which is 3.2.3(i). If A is a countable disjoint union of sets in Σ_1, say $A = \bigcup_1^\infty A_n$, then $\chi_A = \Sigma \chi_{A_n}$ and continuity of U gives $U\chi_A = \Sigma_1^\infty U\chi_{A_n}$. From this we see that $TA = \bigcup_1^\infty TA_n$ and (ii) of Definition 3.2.3 is satisfied.

Next we observe that if $\mu_2(TA) = 0$, then $U\chi_A = 0$ a.e. (μ_2) and so

$$0 = \int |U\chi_A|^p d\mu_2 = \int |\chi_A|^p d\mu_1 = \mu_1(A).$$

Conversely, $\mu_1(A) = 0$ implies $\mu_2(TA) = 0$. We conclude that T is a regular set isomorphism.

Since $\mu_1(\Omega_1) < +\infty$, $\chi_{\Omega_1} = 1 \in L^p$ and we let $h = U\chi_{\Omega_1} = U(1)$. For any $A \in \Sigma_1$, $h(t) = U\chi_A(t) + U\chi_{(\Omega_1 \backslash A)}(t)$, and since the functions on the right have disjoint support, $U\chi_A(t)$ agrees with $h(t)$ whenever $U\chi_A(t)$ is not zero. Therefore,

$$U\chi_A(t) = h(t)\chi_{TA}(t) = h(t)T_1\chi_A(t) \text{ a.e.}$$

and by linearity of U, (38) must hold for every simple function. Since U is an isometry, we have

$$\int_{TA} |h(t)|^p d\mu_2 = \| U\chi_A \|^p = \| \chi_A \|^p = \mu_1(A) = \mu_1(T^{-1}(TA)) = \int_{TA} g(t) d\mu_2,$$

where $g = \frac{d(\mu_1 \circ T^{-1})}{d\mu_2}$. By our earlier remarks, the transformation S given by (37) is an isometry which agrees with U on the simple functions, so $U = S$ and (38) must hold for every function f in $L^p(\Omega_1, \Sigma_1, \mu_1)$. We see also that h satisfies (39).

In the σ-finite case, we assume $\Omega_1 = \bigcup \Omega_n'$ where $\Omega_n' \subset \Omega_{n+1}'$ for each n and follow the standard procedure. For each n, the isometry U induces an isometry U_n on the finite measure space $L^p(\Omega_n', \Sigma_n', \mu_1)$, so that by our previous work there are, for each n, a function h_n and a regular set ismor-phism T_n such that $U_n f_n(t) = h_n(t)T_n' f_n(t)$ for almost all $t \in \Omega_2$, and where f_n denotes the restriction of an $L^p(\Omega_1)$-function to Ω_n', and T_n' denotes the operator induced by T_n. If we let $T(A) = \bigcup_1^\infty T_n(A \cap \Omega_n')$ and $h(t) = h_n(t)$ for $t \in T(\Omega_n')$, with $h(t) = 0$ for $t \in \Omega_2 \backslash T(\Omega_1)$, then both (38) and (39) are satisfied.

The converse statement is simply the statement about the operator S given by (37).   □

The Radon-Nikodym derivative $\frac{d(\mu_1 \circ T^{-1})}{d\mu_2}$ is often called the *conditional expectation* of $|h|^p$ given $T(\Sigma_1)$ and is denoted by $E[|h|^p \mid T(\Sigma_1)]$.

3.2.6. EXAMPLE. *(Grzaslewicz) The function $|h|^p$ need not be measurable with respect to the σ-algebra $T(\Sigma_1)$.*

PROOF. Let Σ_1, Σ_2 denote Lebesgue measurable sets on $[0,1]$ and $[-1,1]$, respectively, and consider the measure spaces $([0,1], \Sigma_1, \lambda)$ and $([-1,1], \Sigma_2, \lambda)$ where λ is Lebesgue measure. For a given r with $1 \leq r < \infty$, let a, b, α, β be positive real numbers such that

$$\alpha\, a^r + \beta\, b^r = 1$$

Define T on Σ_1 by $T(A) = (-\alpha)A \cup \beta A$ and

$$h(t) = \begin{cases} a, & \text{if } -1 \leq t < 0 \\ b, & \text{if } 0 \leq t \leq 1 \end{cases}$$

Then T is a regular set isomorphism of Σ_1 into Σ_2 and the linear operator defined on $L^r[0,1]$ into $L^r[-1,1]$ by

$$Uf(t) = h(t)T_1(f)(t)$$

is an isometry. The conditional expectation $E[|h|^r \mid T(\Sigma_1)]$ is given by $\frac{\alpha a^r + \beta b^r}{\alpha + \beta}$. Note that h is not measurable for $T(\Sigma_1)$.

If for $1 \leq p < q$ we choose α, β so that

$$\alpha a^p + \beta b^p = 1 \text{ and}$$
$$\alpha a^q + \beta b^q = 1,$$

the construction above gives an isometry U on $L^p[0,1] \cap L^q[0,1]$ into $L^p[-1,1] \cap L^q[-1,1]$ which is an isometry for no r different from p or q. The choice of a, b to make things work is a bit delicate for we must have α, β both less than or equal to 1. In fact $\beta < 1$ always holds, but to get $\alpha < 1$ it suffices to choose $1 < b < (\frac{q}{p})^{\frac{1}{q-p}}$ and $a = (1 - \epsilon)$ where $\epsilon < 1 - (\frac{p}{q})\, b^{(q-p)}$. $\square$

3.3. Subspaces of L^p and the Extension Theorem

Following the same line of inquiry as in the previous chapter, we now wish to examine the structure of linear isometries defined on subspaces of L^p spaces. Of crucial importance in this development is the idea of *equimeasurability*.

3.3.1. DEFINITION. *Let $(\Omega_1, \Sigma_1, \mu_1)$ and $(\Omega_2, \Sigma_2, \mu_2)$ be finite measure spaces and let $\mathbb{F}$ denote the scalar field (either $\mathbb{R}$ or $\mathbb{C}$). Given $\mathbb{F}$-valued functions $f_1, f_2, \ldots, f_n$ measurable for Σ_1 and $\mathbb{F}$-valued functions $g_1, g_2, \ldots, g_n$ measurable for Σ_2, we let F be the n-valued function given by $F = (f_1, \ldots, f_n)$ and similarly, $G = (g_1, \ldots, g_n)$. Then we say that F and G are equimeasurable if $\mu_1(F^{-1}(B)) = \mu_2(G^{-1}(B))$ for all Borel sets B in $\mathbb{F}^n$.*

We begin with a theorem concerning equimeasurability that generalizes a
theorem of Rudin. We will follow the work of Hardin in this section although
contributions have been made by a number of authors. (See the notes following
this chapter.)

3.3.2. THEOREM. *(Rudin-Hardin) Let $(\Omega_1, \Sigma_1, \mu_1)$ and $(\Omega_2, \Sigma_2, \mu_2)$ be
finite measure spaces while $F = (f_1, \ldots, f_n)$ and $G = (g_1, \ldots, g_n)$ are n-tuples
of $\mathbb{F}$-valued functions in $L^p(\Omega_1, \Sigma_1, \mu_1)$ and $L^p(\Omega_2, \Sigma_2, \mu_2)$, respectively, where
$0 < p < \infty$ and p is not an even integer. If*

$$(40) \qquad \int_{\Omega_1} |1 + \sum_{j=1}^{n} \lambda_j f_j|^p d\mu_1 = \int_{\Omega_2} |1 + \sum_{j=1}^{n} \lambda_j g_j|^p d\mu_2$$

for all $\lambda_1, \ldots, \lambda_n \in \mathbb{F}$, then F and G are equimeasurable.

If we define α, β on the Borel sets of $\mathbb{F}^n$ by

$$\alpha(B) = \mu_1\{t : (f_1(t), \ldots, f_n(t)) \in B\}$$

and

$$\beta(B) = \mu_2\{t : (g_1(t), \ldots, g_n(t)) \in B\}.$$

Then α, β are Borel measures on $\mathbb{F}^n$. The following theorem is equivalent to
Theorem 3.3.2.

3.3.3. THEOREM. *Let p satisfy $0 < p < \infty$ where p is not an even integer.
If α, β are positive finite Borel measures on $\mathbb{F}^n$ such that*

$$(41) \qquad \int_{\mathbb{F}^n} |1 + \sum_{j=1}^{n} \lambda_j z_j|^p d\alpha(\mathbf{z}) = \int_{\mathbb{F}^n} |1 + \sum_{j=1}^{n} \lambda_j z_j|^p d\beta(\mathbf{z}) < \infty$$

for all $(\lambda_1, \lambda_2, \ldots \lambda_n) \in \mathbb{F}^n$ (where $\mathbf{z} = (z_1, \ldots, z_n) \in \mathbb{F}^n$), then $\alpha = \beta$.

Theorem 3.3.3 is a special case of Theorem 3.3.2. To see that it implies
3.3.2, it suffices to observe that for any simple function σ on $\mathbb{F}^n$,

$$\int_{\mathbb{F}^n} \sigma d\alpha(\mathbf{z}) = \int_{\Omega_1} (\sigma \circ F)(t) d\mu_1(t)$$

where $F = (f_1, f_2, \ldots, f_n)$ and similarly for β, μ_2, and $G = (g_1, g_2, \ldots, g_n)$.
(Here we assume α, β are defined in terms of μ_1, μ_2 as indicated just prior to
the statement of Theorem 3.3.3.)

In the proof of Theorem 3.3.3, use will be made of Fourier Transforms
and in that connection, the space $\mathbb{F}^n$ will be considered as either $\mathbb{R}^n$ or $\mathbb{R}^{2n}$
depending on whether $\mathbb{F}$ is the real or the complex field. We follow the treatment of Fourier Transforms for $\mathbb{R}^n$-valued functions as given by Rudin [274]
and measures as given by Billingsley [34]. In preparation for the proof we
state two lemmas.

For any two finite measures α, β on $\mathbb{F}$, we write $\alpha \sim \beta$ if (41) holds for all $\lambda_1 \in \mathbb{F}$ (where $n = 1$). Let $\mathcal{M}$ denote the set of all bounded, continuous real valued functions f in $L^1(\mathbb{F}, dz)$ such that

$$\int_{\mathbb{F}} f(z) d\alpha(z) = \int_{\mathbb{F}} f(z) d\beta(z)$$

whenever $\alpha \sim \beta$. Here, dz denotes the Lebesgue measure on $\mathbb{F}$.

3.3.4. LEMMA. *If $f \in \mathcal{M}$ and $0 \neq t \in \mathbb{F}$, then $f_t(z) = f(t + z)$ and $f^{(t)}(z) = f(tz)$ are also in $\mathcal{M}$.*

PROOF. For a given $t \neq 0 \in \mathbb{F}^n$, let $\alpha_t(B) = \alpha(t + B)$ for all Borel sets $B \subset \mathbb{F}$. Then for a measurable function f we must have $\int f(z) d\alpha_t(z) = \int f(z - t) d\alpha(z)$ and similarly for β. Now suppose $\alpha \sim \beta$. We want to show that $\alpha_t \sim \beta_t$. Since (41) holds for α, β and $n = 1$, we have

$$\int_{\mathbb{F}} |1 + \lambda z|^p d\alpha_t(z) = \int_{\mathbb{F}} |1 + \lambda(z - t)|^p d\alpha(z) = \int_{\mathbb{F}} |(1 - \lambda t) + \lambda z|^p d_\alpha(z)$$

$$= \int_{\mathbb{F}} |1 - \lambda t + \lambda z|^p d\beta(z) = \int_{\mathbb{F}} |1 + \lambda z|^p d\beta_t(z),$$

except for the case where $(1 - \lambda t) = 0$. For this we need to show that $\int_{\mathbb{F}} |z|^p d\alpha(z) = \int_{\mathbb{F}} |z|^p d\beta(z)$. We sketch this argument for the complex case.

Since $\int_{\mathbb{F}} |q + z|^p d\alpha(z) = \int_{\mathbb{F}} |q + z|^p d\beta(z) < +\infty$ for any nonzero q, and since $|q + z|^p \geq |z|^p$ on $\Re z \geq 0$ for $q > 0$, while $|-q + z|^p \geq |z|^p$ for $\Re z \leq 0$, we conclude that $\int_{\mathbb{F}} |z|^p d\alpha(z)$ and $\int_{\mathbb{F}} |z|^p d\beta(z)$ are finite. Then for $q_n \to 0$, $|q_n + z|^p \leq 2^p(1 + |z|^p)$ for all n, where $q_n \leq 1$ and we get $\int_{\mathbb{F}} |z|^p d\alpha(z) = \int_{\mathbb{F}} |z|^p d\beta(z)$ from the Lebesgue Convergence Theorem.

Now if $f \in \mathcal{M}$, then f_t is bounded, continuous, real valued and in $L^1(\mathbb{F}, dz)$. Furthermore,

$$\int f_t(z) d\alpha(z) = \int f(t + z) d\alpha(z) = \int f(z) d\alpha_t(z)$$

$$= \int f(z) d\beta_t(z) = \int f(t + z) d\beta(z) = \int f_t d\beta(z),$$

so that $f_t \in \mathcal{M}$.

The proof that $f^{(t)} \in \mathcal{M}$ is similar. $\square$

3.3.5. LEMMA. *For p not an even integer, $\mathcal{M}$ has a nontrivial element.*

PROOF. The idea is to show that we can find scalars $\lambda_1, \ldots, \lambda_N$ and $a_1, a_2, \ldots, a_N$ so that $f(z) \equiv \sum_{m=1}^{N} a_m |1 + \lambda_m z|^p$ is bounded, continuous, in L^1, and nontrivial. In that case we will have $f \in \mathcal{M}$ since $\int_{\mathbb{F}} f(z) d\alpha(z) = \int_{\mathbb{F}} f(z) d\beta(z)$ by (41).

For $|z| < 1$, we write

$$(42) \quad |1 + z|^p = [(1 + z)(1 + \bar{z})]^{\frac{p}{2}} = (1 + z)^{\frac{p}{2}}(1 + \bar{z})^{\frac{p}{2}} = \sum_{k, k'=0}^{\infty} b_k b_{k'} z^k \bar{z}^{k'}$$

where $b_k = \binom{\frac{p}{2}}{k}$ and $\bar{z}$ denotes the conjugate of z. (We carry the argument for the case where $\mathbb{F}$ is the complex field.) When $|z| > 1$, we have $|1 + z|^p = |z|^p |1 + z^{-1}|^p$ and so

$$(43) \qquad |1 + z|^p = \sum_{k,k'=0}^{\infty} b_k b_{k'}, |z|^p z^{-k} \bar{z}^{-k'}.$$

Choose N to be an integer greater than $p+3$. For real numbers $a_1, a_2, \ldots, a_n$ form $f(z) = \sum_{m=1}^{n} a_m |1 + mz|^p$. From (43) we get

$$(44) \qquad f(z) = \sum_{k,k'=0}^{\infty} b_k b'_k \left(\sum_{m=1}^{N} a_m m^{p-(k+k')} |z|^p z^{-k} \bar{z}^{-k'} \right)$$

for $|z| > 1$.

Now choose particular $a_1, a_2, \ldots, a_n$ as a nontrivial solution to the system

$$\sum_{m=1}^{n} a_m m^{p-j} = 0 \quad j = 0, 1, 2, \ldots, N - 2$$

and use these in the definition of the function f.

In the power series expansion in (44), the coefficients of $|z|^p z^{-k} \bar{z}^{-k'}$ are zero for $k + k' = 0, 1, \ldots, N - 2$. Now $N - 1 > p + 2$ so we have for $|z| > 2$,

$$|f(z)| \leq \sum_{k+k'>p+2} |b_k| |b_{k'}| \left(\sum_{m=1}^{N} |a_m| \right) |z|^{p-(k+k')}$$

$$\leq constant \; |z|^p \sum_{n>p+2} (n+1) |z|^{-n}$$

$$\leq constant \; |z|^{p-([p]+3)}$$

$$\leq constant \; |z|^{-2}$$

It follows that f is bounded and in L^1, and since f is clearly continuous, we need only show that f is nontrivial. If we restrict f to the real line, note that for p an odd integer, the pth derivative of f will have a discontinuity, while if p is not an integer, the $([p] + 1)^{st}$ derivative will have a discontinuity. Thus f cannot be identically zero. $\qquad \square$

It is important to note here that if p is an even integer, then the function f constructed in the proof above is identically zero. It is only at this one juncture that the hypothesis that p is not an even integer comes into play.

We are finally ready to give the proof of Theorem 3.3.3.

PROOF. The plan is to show that the Fourier transforms $\hat{\alpha}$ and $\hat{\beta}$ are equal, from which $\alpha = \beta$ by the uniqueness theorem for Fourier transforms. Assume $n = 1$. For any $f \in \mathcal{M}$ we have

$$f * \alpha(t) = \int_{\mathbb{F}} f(t - z) d\alpha(z) = \int_{\mathbb{F}} f(t - z) d\beta(z) = f * \beta(t)$$

by Lemma 3.3.4. It follows that

$$\hat{f}(w)\hat{\alpha}(w) = \hat{f}(w)\hat{\beta}(w)$$

for all $w \in \mathbb{F}$ and $f \in \mathcal{M}$. Because of (41), $\hat{\alpha}(0) = \hat{\beta}(0)$ and so we can show $\hat{\alpha} = \hat{\beta}$ if we can show that for each nonzero $w \in \mathbb{F}$, there exists $f \in \mathcal{M}$ with $\hat{f}(w) \neq 0$. Let $w \neq 0$ be fixed in $\mathbb{F}$. Now by Lemma 3.3.5), there is a nontrivial function g in $\mathcal{M}$ which must therefore have the property that $\hat{g}(w_0) \neq 0$ for some $w_0 \neq 0$ in $\mathbb{F}$. If $t = ww_0^{-1}$, then the function $f = g^{(t)} \in \mathcal{M}$ and $\hat{f}(w) = \frac{w_0}{w}\hat{g}(w_0) \neq 0$.

For general n and $\lambda = (\lambda_1, \lambda_2, \ldots, \lambda_n) \in \mathbb{F}^n$, define α_λ and β_λ on the Borel sets of $\mathbb{F}$ by

$$\alpha_\lambda(B) = \alpha(\{z = (z_1, \ldots, z_n) \in \mathbb{F}^n : \sum_{j=1}^{n} \lambda_j z_j \in B\}) \text{ and}$$

$$\beta_\lambda(B) = \beta(\{z = (z_1, \ldots, z_n) \in \mathbb{F}^n : \sum_{j=1}^{n} \lambda_j z_j \in B\}).$$

Then usual consideration for characteristic functions and simple functions leads to

$$\int_{\mathbb{F}^n} f(\sum_{j=1}^{n} \lambda_j z_j) d\alpha(z) = \int_{\mathbb{F}} f(w) d\alpha_\lambda(w), \ f \in L^1(\mathbb{F}, \alpha_\lambda) \text{ and}$$

$$\int_{\mathbb{F}^n} f(\sum_{j=1}^{n} \lambda_j z_j) d\beta(z) = \int_{\mathbb{F}} f(w) d\beta_\lambda(w), f \in L^1(\mathbb{F}, \beta_\lambda).$$

Let $f(w) = |1 + tw|^p$ for a fixed $t \in \mathbb{F}$. Then using (41), and the two equations above, we can conclude that

$$\int_{\mathbb{F}} |1 + tw|^p d\alpha_\lambda(w) = \int_{\mathbb{F}} |1 + tw|^p d\beta_\lambda(w).$$

By what we have already shown for $n = 1$, we must have $\alpha_\lambda = \beta_\lambda$ for all $\lambda \in \mathbb{F}^n$.

Now let $f(w) = e^{-\Re(w)}$. (In the complex case we are considering our measures as defined on $\mathbb{R}^{2n}$, so we write our complex numbers as pairs and interpret the product accordingly.) Thus for any $\lambda \in \mathbb{F}^n$,

$$\hat{\alpha}(\lambda) = \int e^{-i\Re(\sum_{j=1}^{n} \bar{\lambda}_j z_j)} d\alpha(\mathbf{z})$$

$$= \int f(\sum_{j=1}^{n} \bar{\lambda}_j z_j) d\alpha(\mathbf{z}) = \int f(w) d\alpha_{\lambda*}(w)$$

$$= \int f(w) d\beta_{\lambda*}(w) = \int f(\sum_{j=1}^{n} \bar{\lambda}_j z_j) d\beta(\mathbf{z}) = \hat{\beta}(\lambda)$$

where $\lambda^* = (\bar{\lambda}_1, \bar{\lambda}_2, \ldots, \bar{\lambda}_n)$. Hence $\alpha = \beta$ and the proof is complete. $\square$

We now state a corollary which extends Theorem 3.3.3 to the space $\mathbb{F}^\infty$ of all sequences $z = (z_1, z_2, \dots)$ of elements of $\mathbb{F}$, endowed with the product topology.

3.3.6. COROLLARY. *If α, β are positive, finite Borel measures on $\mathbb{F}^\infty$, if $p > 0$ is not an even integer, and if*

$$\int_{\mathbb{F}^\infty} |1 + \sum_{j=1}^N \lambda_j z_j|^p d\alpha(z) = \int_{\mathbb{F}^\infty} |1 + \sum_{j=1}^N \lambda_j z_j|^p d\beta(z) < \infty$$

for all N and $\lambda_1, \dots, \lambda_N \in \mathbb{F}$, then $\alpha = \beta$.

PROOF. For each positive integer N and each Borel set B in $\mathbb{F}^N$, let $\alpha_N(B) = \alpha(\tilde{B})$, $\beta_N(B) = \beta(\tilde{B})$ where $\tilde{B} = \{(z_1, z_2, \dots) \in \mathbb{F}^\infty : (z_1, \dots, z_N) \in B\}$. By Theorem 3.3.3, $\alpha_N = \beta_N$ for all N. Thus α, β agree on the field of all cylinder sets of $\mathbb{F}^\infty$ and hence must agree. □

3.3.7. COROLLARY. *Suppose $p > 0$ is not an even integer. Let M be a subspace of $L^p(\Omega_1 \Sigma_1, \mu_1)$ which contains the constant functions. If $U : M \to L^p(\Omega_2 \Sigma_2, \mu_2)$ is a linear isometry which takes 1 to 1, then $F = (f_1, f_2, \dots)$ and $UF = (Uf_1, Uf_2, \dots)$ are equimeasurable for all $f_1, f_2, \dots$ in M.*

PROOF. Let $\alpha(B) = \mu_1(F^{-1}(B))$ and $\beta(B) = \mu_2((UF)^{-1}(B))$ for all Borel sets B of $\mathbb{F}^\infty$. The result follows from Corollary 3.3.6. □

We now give an example to show that in 3.3.7 (as well as the preceding results), the condition that p is not an even integer cannot be removed.

3.3.8. EXAMPLE. *(Rudin) When p is even, the conclusion of Corollary 3.3.7 can fail.*

PROOF. Let $\Omega_1 = \Omega_2 = (0, \infty)$, $\mathbb{F} = \mathbb{R}$ and for $c \in [-1, 1]$ let $d\lambda_c(t) = t^3 e^{-t}(1 + c \sin t)dt$. For $a \neq b$ with $a, b \in [-1, 1]$, let $\mu_1 = \lambda_a$ and $\mu_2 = \lambda_b$, and let p be any even positive integer. Let M be the algebra of polynomials $c_0 + c_1 t^4 + \dots + c_m t^{4m}$, $c_j \in \mathbb{R}$ and let $U : M \to M$ be the identity map. Then $f(t) = t^4$ is in M, but f and Uf are not equimeasurable. This fact rests on the identity

$$\int_0^\infty t^{s-1} e^{-t} \sin t\, dt = \Gamma(s) 2^{\frac{-s}{2}} \sin \frac{s\pi}{4}$$

which can be proved by replacing $\sin t$ with $\frac{e^{it} - e^{-it}}{2i}$ and applying Cauchy's Theorem. Then it can be shown that

$$\int_0^\infty |g|^{2k} d\mu_1 = \int_0^\infty |g|^{2k} d\mu_2$$

for every $g \in M$ and every $k = 1, 2, \dots$. However,

$$\int_0^\infty f^{\frac{1}{2}} d\mu_1 \neq \int_0^\infty f^{\frac{1}{2}} d\mu_2$$

which is enough to conclude that f and $Uf = f$ are not equimeasurable. (If f and g are equimeasurable, then $\int(h \circ f)d\mu_1 = \int(h \circ g)d\mu_2$ for every Borel measurable function h.)   □

In preparation for the main theorems in this section, we introduce some useful and special notation. If S is a collection of subsets of a set Ω, then $\varsigma(S)$ will denote the smallest σ-algebra on Ω which contains S; i.e., the σ-algebra generated by S. If M is a collection of $\mathbb{F}$-valued functions on Ω, then $\varsigma(M)$ will denote the smallest σ-algebra on Ω for which each function in M is measurable.

3.3.9. THEOREM. *Let* $(\Omega_1, \Sigma_1, \mu_1)$, $(\Omega_2, \Sigma_2, \mu_2)$ *be finite measure spaces and* M *a subspace of* $L^p(\Omega_1, \Sigma_1, \mu_1)$ *which contains the constant functions. Suppose that* $0 < p < \infty$ *and* p *is not an even integer. If* U *is a linear isometry from* M *into* $L^p(\Omega_2, \Sigma_2, \mu_2)$ *for which* $U1 = 1$, *then* U *has the form*

$$Uf = T_1 f$$

where T_1 *is induced by a measure preserving regular set isomorphism of* $\sigma(M)$ *onto* $\sigma(U(M))$. *Further, the induced* T_1 *is the unique extension of* U *to an isometry from* $L^p(\sigma(M))$ *onto* $L^p(\sigma(U(M)))$.

PROOF. Suppose $C = \{f_1, f_2, \ldots\}$ is a countable collection of functions in M. Then $\sigma(C) = \{F^{-1}(B) : B \text{ a Borel set in } \mathbb{F}^\infty\}$, where $F(t) = (f_1(t), f_2(t), \ldots)$. Hence for any $E \in \sigma(C)$, there is a Borel set $E' \subseteq \mathbb{F}^\infty$ so that $E = F^{-1}(E')$. We define a set map $T_C : \sigma(C) \to \sigma(U(C))$ by

$$T_C(E) = (UF)^{-1}(E'),$$

where $UF(t) = (Uf_1(t), Uf_2(t), \ldots)$. If $E = F^{-1}(E') = F^{-1}(E'')$, then

$$\mu_1(F^{-1}(E' \Delta E'')) = \mu_1(E \Delta E) = 0$$

and by Corollary 3.3.7,

$$\mu_2((UF)^{-1}(E') \Delta (UF)^{-1}(E'')) = \mu_2((UF)^{-1}(E' \Delta E'')) = 0$$

as well. Hence T_C is well defined.

Furthermore, we also see by Corollary 3.3.7 that for any $E \in \sigma(C)$,

$$\mu_1(E) = \mu_1(F^{-1}(E')) = \mu_2((UF)^{-1}(E')) = \mu_2(T_C(E)).$$

Thus T_C is measure preserving and it is straightforward to verify that T_C is a regular set isomorphism on $\sigma(C)$ to $\sigma(U(C))$. The explicit description of elements of $\sigma(U(C))$ shows that T_C must be surjective. Now $T_C(\Omega_1) = \Omega_2$, so that the operator induced by T_C (also denoted by T_C) must map 1 to 1, and is an isometry on the simple functions determined by $\sigma(C)$. It follows that T_C defines an isometry on $L^p(\sigma(C))$ which is onto $L^p(\sigma(U(C)))$ (since the set map T_C is surjective).

By part (v)c of Remark 3.2.4, we see that for a given j,

$$(T_C f_j)^{-1}(E) = T_C(f_j^{-1}(E))$$

for every Borel set E in $\mathbb{F}$. For such a Borel set E and a given j, let $\tilde{E}_j$ denote the Borel set in $\mathbb{F}^\infty$ which is a product of Borel sets in $\mathbb{F}$ where each factor is $\mathbb{F}$ except in the jth position where the entry is E. Then

$$(T_C f_j)^{-1}(E) = T_C(f_j^{-1}(E)) = T_C(F^{-1}(\tilde{E}_j)) = (UF)^{-1}(\tilde{E}_j) = (U f_j)^{-1}(E).$$

Since $(T_C f_j)^{-1}(E) = (U f_j)^{-1}(E)$ for all Borel sets E, we conclude that $T_C f_j = U f_j$ for each j. Hence we can see that T_C is an isometry on $L^p(\sigma(C))$ into $L^p(\Sigma_2)$ which agrees with U on the closed linear span of $\{1, f_1, f_2, \ldots\}$.

At this point we wish to show that T_C is the only such isometry. Assume that V is another isometry on $L^p(\sigma(C))$ which agrees with U on the closed linear span of $\{1, f_1, f_2, \ldots\}$. Suppose E is any set in $\sigma(U(C))$ and $f \in L^p(\sigma(C))$. For any Borel set A in $\mathbb{F}$, we have $E = (UF)^{-1}(E')$ for a Borel set E' and

$$
\begin{aligned}
\mu_2((T_C f)^{-1}(A) \cap E) &= \mu_2(\{t : (T_C f(t), U f_1(t), \ldots) \in A \times E'\}) \\
&= \mu_1(\{t : (f(t), f_1(t), \ldots) \in A \times E'\}) \\
&= \mu_2(\{t : (V f(t), V f_1(t), \ldots) \in A \times E'\}) \\
&= \mu_2(\{t : (V f(t), U f_1(t), \ldots) \in A \times E'\}) \\
&= \mu_2((V f)^{-1}(A) \cap E),
\end{aligned}
$$

where the second and third equalities above hold because of Corollary 3.3.7. Since $(T_C f)^{-1}(A)$ and $(U f)^{-1}(A)$ are both in $\sigma(U(C))$, the above equality for all $E \in \sigma(U(C))$ implies that $\mu_2((T_C f)^{-1}(A)\Delta(V f)^{-1}(A)) = 0$ and we have $(T_C f)^{-1}(A) = (V f)^{-1}(A)$ for all Borel sets A in $\mathbb{F}$. We conclude that $T_C f = V f$.

To finish the proof we must show how to define the operator T on $L^p(\sigma(M))$. Given $f \in L^p(\sigma(M))$, it follows by a theorem given by Doob [87, p.604] that there is a countable set $C = \{f_1, f_2, \ldots\} \subset M$ such that $f \in \sigma(C)$. We define T by

$$Tf = T_C f$$

where T_C is the unique operator defined above. If C' is another countable collection for which $f \in \sigma(C')$, the uniqueness of the operator $T_{C \cup C'}$ leads to the conclusion that $T_C(f) = T_{C'}(f)$ and T is well defined. If S is another such isometry which agrees with U on M, then it follows that for any $C = \{f_1, f_2, \ldots\} \subset M$, S restricted to $L^p(\sigma(C))$ must be T_C, and so $S = T$. Finally, by Lamperti's Theorem 3.2.5, the operator T is induced by a regular set isomorphism which necessarily maps $\sigma(M)$ onto $\sigma(U(M))$, so that T maps $L^p(\sigma(M))$ onto $L^p(\sigma(U(M))$ as advertised.

□

Our goal now is to relax the assumptions of Theorem 3.3.9, and characterize isometries on an arbitrary closed subspace of $L^p(\mu_1)$. Again we can obtain an extension to an explicitly defined subspace. For this we will need some new definitions.

3.3.10. DEFINITION. *Let (Ω, Σ, μ) denote a measure space.*

(i) *An element f of a collection M of measurable functions on Ω is said to have* full support *in M if $f \in M$ and $supp(g) \backslash supp(f)$ is a null set for any $g \in M$.*

(ii) *The* ratio σ-algebra *generated by M is the σ-algebra $\rho(M) = \sigma\{(f(\frac{1}{g}) : f, g \in M\}$. We let $\mathcal{R}(M)$ denote the set of all $\rho(M)$-measurable functions on Ω.*

(iii) *If f_o is an element of full support in M, we let $\tilde{\rho}(M)$ denote the elements of $\rho(M)$ intersected with $supp(f_o)$, and $\tilde{\mathcal{R}}(M)$ denotes the set of $\tilde{\rho}(M)$-measurable functions.*

Note that $\tilde{\rho}(M)$ and $\tilde{\mathcal{R}}(M)$ are independent of the choice of f_o, $\tilde{\rho}(M) \subseteq \rho(M) \subseteq \sigma(M)$, and $\tilde{\mathcal{R}}(M) \cdot M = \mathcal{R}(M) \cdot M$ since $supp(f_o)$ is $\rho(M)$-measurable.

Our first task is to show that elements of full support actually exist.

3.3.11. LEMMA. *(Hardin) Let (Ω, Σ, μ) denote a σ-finite measure space and suppose M is a closed subspace of $L^p(\mu)$ for some p, $0 < p < \infty$. Then there exists an element of full support in M.*

PROOF. Since μ is σ-finite there is a finite measure $\bar{\mu}$ which is equivalent to μ (i.e., each of $\bar{\mu}$ and μ are absolutely continuous with respect to the other). For a finite subset B of M let $s(B) = \bar{\mu}(\bigcup_{f \in B} supp(f))$ and let $s = sup\{s(B) : B$ is a finite subset of $M\}$. Let $\{B_n\}$ be a sequence of finite subsets of M such that $s(B_n) \leq s(B_{n+1})$ for each n and $s(B_n) \to s$. It can be shown that for the countable set $C = \bigcup_1^n B_n$,

$$\bar{\mu}(supp(g) \backslash \bigcup_{f \in C} supp(f)) = 0$$

for any $g \in M$. By the equivalence of μ and $\bar{\mu}$, we get that

$$\mu(supp(g) \backslash \bigcup_{n=1}^{\infty} supp(g_n)) = 0$$

for any $g \in M$, where $C = \{g_1, g_2, \ldots\}$. Now let $f_n = (2^n \|g_n\|^p)^{\frac{-1}{p}} g_n$ for each n so that $\sum_1^{\infty} \|f_n\|^p < +\infty$. It follows that $\sup_n\{|f_n(t)|\} < +\infty$ almost everywhere (μ).

For each n, let $r_n(w)$ be the linear function defined on $[0, 1]$ with $r_n(0) = -2^{-n}$ and $r_n(1) = 2^{-n}$. Define g on $[0, 1] \times \Omega$ by

$$g(w, t) = \sum_{n=1}^{\infty} r_n(w) f_n(t)$$

where the series converges and zero elsewhere. Then the series converges almost everywhere on $[0, 1] \times \Omega$ with respect to the product measure $\lambda \times \mu$ where λ is Lebesgue measure on $[0, 1]$.

Let

$$S = \bigcup_{n=1}^{\infty} supp(f_n),$$

$$A = \{(w, t) \in [0, 1] \times S : g(w, t) = 0\},$$

$$A_t = \{w \in [0, 1] : g(w, t) = 0\}, \quad and$$

$$A^w = \{t \in S : g(w, t) = 0\}.$$

Then $\sum_1^{\infty} r_n(w)f_n(t)$ converges for almost all t in S and for such t, the measure $\lambda(g^{-1}(\cdot, t))$ is absolutely continuous with respect to Lebesgue measure on $\mathbb{F}$, so that we must have $\lambda(A_t) = \lambda(g^{-1}(\cdot, t))(\{0\}) = 0$. It now follows from Tonelli's Theorem that

$$0 = \int_S \lambda(A_t)d\mu(t) = \int_A d(\lambda \times \mu) = \int_{[0,1]} \mu(A^w)d\lambda(w)$$

from which we may conclude that $\mu(A^w) = 0$ for almost all w in $[0, 1]$. Hence, we may choose $w_o \in [0, 1]$ for which $\mu(A^{w_o}) = 0$ and let

$$f_o(t) = \sum_1^{\infty} r_n(w_o)f_n(t)$$

if the series converges, and zero otherwise. This function $f_o \in M$ and has full support in M. $\qquad \square$

3.3.12. LEMMA. *Suppose $\{f_n\}$ and $\{g_n\}$ are sequences in $L^p(\mu)$ and $L^q(\nu)$, respectively, where $0 < p, q < \infty$, and μ, ν are σ-finite. Let M, N be the closed linear spans of the sequences $\{f_n\}$ and $\{g_n\}$, respectively. Then there exists a sequence $\{a_n\}$ of real numbers such that the sums $f_o = \sum a_n f_n$ and $g_o = \sum a_n g_n$ converge a.e. in $L^p(\mu)$ and $L^q(\nu)$, respectively, and define functions of full support in M and N, respectively.*

PROOF. We may assume that $\sum \|f_n\|_p^p$ and $\sum \|g_n\|_q^q$ both converge. Define the functions $\{r_n\}$ as in the proof of the previous lemma and obtain w_o as in that proof so that $\mu(A^{w_o})$ and $\nu((A')^{w_o}) = 0$. Let $a_n = r_n(w_o)$ for each n. $\qquad \square$

Now we show that an isometry must take functions of full support to functions of full support.

3.3.13. LEMMA. *Let M be a closed subspace of $L^p(\Omega_1, \Sigma_1, \mu_1)$ and U a linear isometry of M into $L^p(\Omega_2, \Sigma_2, \mu_2)$. If $0 < p < \infty$ and p is not an even integer, and if f_o has full support in M, then Uf_o has full support in $U(M)$.*

PROOF. Let $g_1 = Uf_1$ be any element of $U(M)$ and let $g_o = Uf_o$, where f_o is an element of full support in M. By Lemma 3.3.12 we may find constants a_o, a_1 such that the functions $f_2 = a_o f_o + a_1 f_1$ and $g_2 = a_o g_o + a_1 g_1$ have full support in $sp\{f_o, f_1\}$ and $sp\{g_o, g_1\}$, respectively. Since U is an isometry, we have

$$\int \left| 1 + \lambda f_o \frac{1}{f_2} \right|^p |f_2|^p d\mu_1 = \int |f_2 + \lambda f_o|^p d\mu_1$$

$$= \int |g_2 + \lambda g_o|^p d\mu_2$$

$$= \int \left| 1 + \lambda g_o \frac{1}{g_2} \right|^p |g_2|^p d\mu_2$$

for all $\lambda \in \mathbb{F}$. Note that the products $f_o \left(\frac{1}{f_2} \right)$ and $g_o \left(\frac{1}{g_2} \right)$ are finite a.e. because $supp(f_o) \subseteq supp(f_2)$ and $supp(g_o) \subseteq supp(g_2)$ (except possibly for null sets).

If we let $d\nu_1 = |f_2|^p d\mu_1$ and $d\nu_2 = |g_2|^p d\mu_2$, the string of equalities above shows that the linear transformation which takes 1 to 1 and $f_o \left(\frac{1}{f_2} \right)$ to $g_o \left(\frac{1}{g_2} \right)$ is an isometry on $sp\{1, f_o \left(\frac{1}{f_2} \right)\}$ as a subspace of $L^p(\nu_1)$ into $L^p(\nu_2)$.

Using Corollary 3.3.7 once again, we see that the functions $g_o \left(\frac{1}{g_2} \right)$ and $f_o \left(\frac{1}{f_2} \right)$ are equimeasurable for ν_2 and ν_1, so that

$$\nu_2(\{t : \left(g_o \left(\frac{1}{g_2} \right) \right)(t) = 0\}) = \nu_1(\{t : \left(f_o \frac{1}{f_2} \right)(t) = 0\}) = 0,$$

where the last equality comes from the definition of ν_1 and the fact that $supp(f_2) \subseteq supp(f_o)$. From this it follows that $\mu_2(\{t : \left(g_o \frac{1}{g_2} \right)(t) = 0\} \cap supp(g_2)) = 0$, so that, except for null sets,

$$supp(g_1) \subseteq supp(g_2) = supp(g_o).$$

$\square$

We are finally ready to state and prove the main result of this section.

3.3.14. THEOREM. *(Hardin) Suppose U is a linear isometry from a closed subspace M of $L^p(\Omega_1, \Sigma_1, \mu_1)$ into $L^p(\Omega_2, \Sigma_2, \mu_2)$ where $0 < p < \infty$ and p is not an even integer. Then U has a unique extension to an isometry $\bar{U}$ on $\mathcal{R}(M) \cdot M \cap L^p(\mu_1)$ which maps onto $\mathcal{R}(U(M)) \cdot U(M) \cap L^p(\mu_2)$ and satisfies*

(45) $$\bar{U}(rf) = (T_1 r)(Uf)$$

for any r in $\tilde{\mathcal{R}}(M)$ and $f \in M$. The operator T_1 is induced by a regular set isomorphism T of $\tilde{\rho}(M)$ onto $\tilde{\rho}(U(M))$ and

(46) $$E \left(|Uf|^p \Big| \tilde{\rho}(U(M)) \right) = \frac{d(\mu_f \circ T^{-1})}{d\mu_1}$$

for all $f \in M$ where $d\mu_f = |f|^p d\mu_1$.

PROOF. By Lemma 3.3.11, there exists $f_o \in M$ with full support in M. We will assume that $supp(f_o) = \Omega_1$, which is clearly possible without any loss in generality. With this assumption we have $\tilde{\rho}(M) = \rho(M)$ and $\widetilde{\mathcal{R}}(M) = \mathcal{R}(M)$. Let $M_o = \{f\frac{1}{f_o} : f \in M\}$. We observe that for any $f, g \in M$, $f \cdot \frac{1}{g} = \left(f\frac{1}{f_o}\right)\left(g\frac{1}{f_o}\right)^{-1}$ from which it follows that $\sigma(M_o) = \rho(M)$.

If $rf \in \mathcal{R}(M) \cdot M \cap L^p(\mu_1)$, then

$$rf = \left(rf \cdot \frac{1}{f_o}\right) f_o = r_o f_o$$

where r_o is $\rho(M)$-measurable. Hence,

$$\mathcal{R}(M) \cdot M \cap L^p(\mu_1) = S(f_o) \equiv \{r_o f_o \in L^p(\mu_1) : r_o \text{ is } \sigma(M_o) - measurable\}.$$

Let $d\mu_o = |f_o|^p d\mu_1$ and $d\nu_o = |Uf_o|^p d\mu_2$. We define $G_o = \{Uf \cdot \frac{1}{Uf_o} : f \in M\}$ and note that by Lemma 3.3.13, Uf_o has full support in $U(M)$ so that the functions in G_o are finite almost everywhere. Now M_o and G_o may be regarded as subspaces of $L^p(\mu_o)$ and $L^p(\nu_o)$, respectively, and each contains the 1-function. If we define U_o on M_o by $U_o\left(f \cdot \frac{1}{f_o}\right) = Uf\left(\frac{1}{Uf_o}\right)$, we can see that U_o is an isometry of M_o onto G_o and $U_o1 = 1$. We apply Theorem 3.3.9 to obtain a unique extension of U_o to an isometry T_1 from $L^p(\sigma(M_o), \mu_o)$ onto $L^p(\sigma(G_o), \nu_o)$ which is induced by a measure preserving regular set isomorphism T of $\sigma(M_o) = \rho(M)$ onto $\sigma(G_o) = \tilde{\rho}(U(M))$.

Next we define operators $V_o : S(f_o) \to L^p(\sigma(M_o), \mu_o)$ by

$$V_o f = f \cdot \left(\frac{1}{f_o}\right)$$

and $W_o : L^p(\sigma(G_o), \nu_o) \to L^p(\mu_2)$ by

$$W_o f = fUf_o.$$

It is straightforward to show that V_o and W_o are surjective isometries and that $\bar{U} = W_o T_1 V_o$ is an isometry from $S(f_o)$ into $L^p(\mu_2)$ which extends U. If $r \in \mathcal{R}(M)$ and $f \in M$ we have, letting $r_o = rf \cdot \frac{1}{f_o}$, that

$$\bar{U}(rf) = W_o T_1 V_o(r_o f_o) = W_o(T_1 r_o) = T_1(r_o)U(f_o)$$

$$= T_1(r)\left(T_1\left(f \cdot \frac{1}{f_o}\right)\right)Uf_o \qquad \text{(by 3.2.4,(v)d)}$$

$$= T_1(r)\left(U_o\left(f\frac{1}{f_o}\right)\right)Uf_o$$

$$= T_1(r)U(f)$$

which establishes (45). Note that $T_1(r) \in \mathcal{R}(U(M))$ and $\bar{U}$ maps $\mathcal{R}(M) \cdot M \cap L^p(\mu_1)$ onto $\mathcal{R}(U(M)) \cdot U(M) \cap L^p(\mu_2)$.

Now let us assume that $V : S(f_o) \to L^p(\mu_2)$ is another isometry which extends U to $S(f_o)$. Then $W_o^{-1}VV_o^{-1}$ is an isometry from $L^p(\sigma(M_o), \mu_1)$ onto

$L^p(\sigma(G_o), \nu_o)$ which extends U_o, and so must equal T_1 since T_1 is the only such isometry by Theorem 3.3.9. Hence, $V = W_o T_1 V_o = \bar{U}$, and $\bar{U}$ is unique.

To complete the proof we wish to establish (46). Let $A \in \tilde{\rho}(U(M))$. Then

$$(\mu_f \circ T^{-1})(A) = \mu_f(T^{-1}(A)) = \int_{T^{-1}(A)} |f|^p d\mu_1$$

$$= \int |\chi_{T^{-1}(A)} f|^p d\mu_1 = \int |\bar{U}(\chi_{T^{-1}(A)} f)|^p d\mu_2$$

$$= \int |T_1(\chi_{T^{-1}(A)}) U f|^p d\mu_2 \qquad \text{(by 45)}$$

$$= \int |\chi_A U f|^p d\mu_2$$

$$= \int_A |U f|^p d\mu_2$$

from which (46) follows.

$\square$

We conclude the section by stating one last corollary.

3.3.15. COROLLARY. *(Hardin) Let $L^p(\mu)$ and $L^p(\nu)$ be as in the statement of (3.3.14) with $0 < p < \infty$, p not an even integer. If M is a closed subspace of $L^p(\mu)$ which contains 1 and U is an isometry of M into $L^p(\nu)$, then U has a unique extension to an isometry $\bar{U}$ from $L^p(\sigma(M), \mu)$ into $L^p(\nu)$ such that*

$$\bar{U} f = T_1 f \cdot U 1$$

where T_1 is induced by a regular set isomorphism T on $\sigma(M)$ such that

$$E\left(|U1|^p \big| T(\sigma(M))\right) = \frac{d(\mu \circ T^{-1})}{d\nu}.$$

3.3.16. REMARKS. *A very interesting application of the above Corollary shows that any isometry U on the two dimensional subspace $sp\{1, t\} \subset L^p[0,1]$ to a space $L^p(\nu)$ has a unique extension to an isometry $\bar{U} : L^p[0,1] \to L^p(\nu)$, where $0 < p$ and p is not an even integer.*

3.4. Bochner Kernels

It is not difficult to verify that the operator U defined on $L^2[0, 2\pi]$ by

$$U f(t) = f(t) + \frac{1}{2\pi}\left[\left(\frac{1}{\sqrt{2}} - 1\right) + \frac{1}{\sqrt{2}} e^{it}\right] \int_0^{2\pi} f(t) dt$$

$$+ \frac{1}{2\pi}\left[\left(\frac{1}{\sqrt{2}} - 1\right) e^{it} - \frac{1}{\sqrt{2}}\right] \int_0^{2\pi} f(t) e^{-it} dt$$

is unitary. This operator is not in the "canonical" form of (38), and we see why it has been necessary to assume $p \neq 2$ in the results of the previous sections. Indeed, the operator U does not satisfy the disjoint support property and such operators cannot be in canonical form. The action can be described, however, in terms of integrable kernels, and we examine this point of view in this section.

The inspiration for the work here comes from Plancherel, whose famous theorem showed how to extend the Fourier Transform from $L^1 \cap L^2$ to all of L^2. The formulas

$$g(t) = \frac{1}{\sqrt{2\pi}} \frac{d}{dt} \int_{-\infty}^{\infty} \frac{e^{-ist} - 1}{-is} f(s) ds$$

$$f(t) = \frac{1}{\sqrt{2\pi}} \frac{d}{dt} \int_{-\infty}^{\infty} \frac{e^{ist} - 1}{is} g(s) ds$$

define a unitary transformation of $L^2(-\infty, \infty)$ and its inverse. The formulas can also be written in the classical form

$$g(t) = \frac{1}{\sqrt{2\pi}} \int_{-\infty}^{\infty} e^{-ist} f(s) ds, \quad f(t) = \frac{1}{\sqrt{2\pi}} \int_{-\infty}^{\infty} e^{ist} g(s) ds$$

if we interpret the integrals to be the limits (in the L^2-norm) of the integral from $-n$ to n as $n \to \infty$. We now state and prove a theorem of Bochner which extends this kernel representation to all unitary operators on L^2.

3.4.1. THEOREM. *(Bochner) Let U denote a unitary operator on $L^2(a, b)$ where $-\infty \le a < b \le \infty$. Then there are complex valued functions $K(s,t)$ and $H(s,t)$ defined on $(a,b) \times (a,b)$ such that for $g = Uf$, we have*

$$(47) \qquad \int_0^s g(t) dt = \int_a^b \overline{K(s,t)} f(t) dt \ \text{ and } \int_0^s f(t) dt = \int_a^b \overline{H(s,t)} g(t) dt;$$

where

(i) $\int_a^b \overline{K(s,t)} K(u,t) dt = \begin{cases} min\{|s|, |u|\} & \text{if } su \ge 0 \\ 0 & \text{if } su \le 0 \end{cases}$

(ii) $\int_a^b \overline{H(s,t)} H(u,t) dt = \begin{cases} min\{|s|, |u|\} & \text{if } su \ge 0 \\ 0 & \text{if } su \le 0 \end{cases}$

(iii) $\int_0^u K(s,t) dt = \int_0^s \overline{H(u,t)} dt$.

Conversely, for a pair H, K of kernels satisfying (i), (ii), (iii) above, the formulas in (47) define a unitary transformation and its inverse.

PROOF. If U is unitary on L^2, define H and K by

$$H(s,t) = Ue_s(t) \quad \text{and} \quad K(s,t) = U^{-1} e_s(t)$$

where e_s is the characteristic function of $[0, s]$ for $s \ge 0$ and $e_s = -\chi_{[s,0]}$ for $s < 0$. Now, if $g = Uf$ we have

$$\langle g, e_s \rangle = \langle Uf, e_s \rangle = \langle f, U^{-1} e_s \rangle,$$

while

$$\langle f, e_s \rangle = \langle U^{-1} g, e_s \rangle = \langle g, Ue_s \rangle.$$

These equations give the formulas in (47).

Now if we let $f = U^{-1}e_u = K(u,t)$, then $g = e_u$ and

$$\int_0^s g(t)dt = \int_a^b \overline{K(s,t)}f(t)dt = \int_a^b \overline{K(s,t)}K(u,t)dt$$

so that (i) holds since $\int_0^s e_u(t)dt = min(|s|,|u|)$ if $su \geq 0$ and zero if $su < 0$. Similarly (ii) follows by letting $f = e_u$, and the equalities

$$\int_0^u K(s,t)dt = \int_a^b K(s,t)e_u(t)dt = \int_0^s Ue_u(t)dt = \int_0^s \overline{H(u,t)}dt$$

establish (iii).

For the converse, we define transformations U, V by letting $Ue_s(t) = H(s,t)$, and $Ve_s(t) = K(s,t)$. Conditions (i), (ii), and (iii) lead to the equations

$$\langle Ve_s, Ve_u \rangle = \langle e_s, e_u \rangle = \langle Ue_s, Ue_u \rangle \quad and$$
$$\langle Ve_s, e_u \rangle = \langle e_s, Ue_u \rangle.$$

Since a step function on (a,b) can be written as a finite linear combination of the functions e_s, we can extend U, V to the step functions in such a way that

$$(48) \qquad \langle Vf, Vg \rangle = \langle f, g \rangle, \quad \langle Uf, Ug \rangle = \langle f, g \rangle, \quad and \quad \langle Vf, g \rangle = \langle f, Ug \rangle.$$

By the density of the step functions in L^2, we can get U, V extended to all of L^2 so that the equations (48) hold for all $f, g \in L^2$. It follows that U and V are isometries with $V = U^* = U^{-1}$.

$\square$

We close this section with a generalization of Bochner's Theorem that treats operators from L^p to L^q. Let $(\Omega_1, \Sigma_1, \mu_1)$ and $(\Omega_2, \Sigma_2, \mu_2)$ be measure spaces and let $1 \leq p < \infty$, $1 \leq q < \infty$. A subring $\mathfrak{A}_0$ of elements of Σ_1 will be called a *fundamental family* for $L^p(\mu_1)$ if the elements of $\mathfrak{A}_0$ have finite measure and the finite linear combinations of characteristic functions of elements in $\mathfrak{A}_0$ are dense in L^p.

For q' such that $\frac{1}{q} + \frac{1}{q'} = 1$, let $\mathfrak{A}_0, \mathfrak{B}_0$ denote fixed fundamental families for $L^p(\Omega_1, \Sigma_1, \mu_1)$, $L^{q'}(\Omega_2, \Sigma_2, \mu_2)$, respectively.

3.4.2. DEFINITION. *Let $H = \{H_B : B \in \mathfrak{B}_0\}$ and $K = \{K_A : A \in \mathfrak{A}_0\}$ denote families of functions on Ω_1 and Ω_2, respectively. The pair (H, K) is called a* Bochner Kernel of type (p,q) *if*

(i) $H_B \in L^{p'}(\mu_1)$ *whenever $B \in \mathfrak{B}_0$ and $K_A \in L^q(\mu_2)$ whenever $A \in \mathfrak{A}_0$;*

(ii) $\|K_A\|_q^p = \mu_1(A)$ *for every $A \in \mathfrak{A}_0$;*

(iii) *If $\{A_i\}_{i=1}^n \subset \mathfrak{A}_0$ with $\mu_1(A_i \cap A_j) = 0$ for $i \neq j$, and if $\{\lambda_i\}_{i=1}^n$ are scalars, then*

$$\|\sum_1^n \lambda_i K_{A_i}\|_q^p = \sum_1^n |\lambda_i|^p \|K_{A_i}\|_q^p;$$

(iv) *If $A \in \mathfrak{A}_0$, $B \in \mathfrak{B}_0$, then $\int_A H_B d\mu_1 = \int_B \overline{K_A} d\mu_2$.*

This definition includes the kernels as defined in Bochner's Theorem where the fundamental families would consist of finite open intervals.

For purposes of stating and proving the next theorem, let us denote by $\langle \cdot, \cdot \rangle_p$ the sesquilinear form defined on $L^p \times L^{p'}$ by $\langle f, g \rangle_p = \int f\bar{g}$. Also, given a bounded operator V from L^p into L^q, we will let V^* be the operator on $L^{q'}$ to $L^{p'}$ defined by

$$\langle Vf, g \rangle_q = \langle f, V^*g \rangle_p.$$

Then V^* is an isometry when restricted to the dual of the range of V and corresponds to the usual adjoint operator when $p = 2$.

3.4.3. THEOREM. *(Merlo) For each Bochner kernel (H, K) of type p, q there is an isometry $U : L^p(\Omega_1, \Sigma_1, \mu_1) \to L^q(\Omega_2, \Sigma_2, \mu_2)$ so that for $f \in L^p$ and $B \in \mathfrak{B}_0$,*

$$(49) \qquad \int_B Ufd\mu_2 = \int f\overline{H_B}d\mu_1,$$

and if $g \in L^{q'}$, $A \in \mathfrak{A}_0$, then

$$(50) \qquad \int_A U^*gd\mu_1 = \int g\overline{K_A}d\mu_2.$$

Conversely, an isometry U from L^p to L^q determines a Bochner kernel (H, K) so that the above equations hold.

PROOF. Let (H, K) denote a Bochner kernel of type p, q. For each $A \in \mathfrak{A}_0$, define $U\chi_A = K_A$. By 3.4.2(ii),

$$\|U\chi_A\|_q = \|K_A\|_q = \mu_1(A)^{\frac{1}{p}} = \|\chi_A\|_p.$$

Let φ be a simple function, $\varphi = \sum_{i=1}^n \lambda_i \chi_{A_i}$, where $A_i \in \mathfrak{A}_0$ for each i and $A_i \cap A_j = \emptyset$ for $i \neq j$. Then by 3.4.2(iii), we have

$$\|U\varphi\|_q^p = \|\sum_{i=1}^n \lambda_i K_{A_i}\|_q^p = \sum_1^n |\lambda_i|^p \|K_{A_i}\|_q^p = \sum_1^n |\lambda_i|^p \|\chi_{A_i}\|_p^p$$

$$= \|\sum_1^n \lambda_i \chi_{A_i}\|_p^p = \|\varphi\|_p^p.$$

Since any finite linear combination of characteristic functions of elements of $\mathfrak{A}_0$ can be written in the above form, we have that U is an isometry on this dense class. We may therefore extend U to an isometry on all of L^p.

Now for $A \in \mathfrak{A}_0$ and $B \in \mathfrak{B}_0$, we see that

$$\langle U\chi_A, \chi_B \rangle_q = \int_B K_A d\mu_2 = \int_A \overline{H_B}d\mu_1 = \langle \chi_A, H_B \rangle_p$$

where the second inequality follows from 3.4.2(iv). By linearity and density, we can get

$$\langle Uf, \chi_B \rangle = \langle f, \chi_B \rangle$$

for all $f \in L^p$ which establishes (49). Furthermore,

$$\int_A U^* \chi_B d\mu_1 = \overline{\langle \chi_A, U^* \chi_B \rangle_p} = \overline{\langle U \chi_A, \chi_B \rangle_q} = \int_B \overline{K_A} d\mu_2.$$

Once again we extend to simple functions which are linear combinations of elements of $\mathfrak{B}_0$ and then by density so that

$$\int_A U^* g d\mu_1 = \int g \overline{K_A} d\mu_2$$

for all $g \in L^{q'}$ and we have (50).

For the converse, suppose U is a linear isometry from L^p into L^q. For each $A \in \mathfrak{A}_0$, $B \in \mathfrak{B}_0$ define $K_A = U \chi_A$, $H_B = U^* \chi_B$. Then $K_A \in L^q$, $H_B \in L^{p'}$ and for $f \in L^p$ we have

$$\int_B U f d\mu_2 = \langle U f, \chi_B \rangle_q = \langle f, U^* \chi_B \rangle_p = \langle f, H_B \rangle_p = \int f \overline{H_B} d\mu_1$$

so that (49) holds. Similarly we get (50) from

$$\int_A U^* g d\mu_1 = \overline{\langle \chi_A, U^* g \rangle_p} = \overline{\langle U \chi_A, g \rangle_q} = \int g \overline{K_A} d\mu_2.$$

For any $A \in \mathfrak{A}_0$,

$$\|K_A\|_q = \|U \chi_A\|_q = \|\chi_A\|_p = (\mu_1(A))^{\frac{1}{p}}$$

which is 3.4.2(ii). For a disjoint collection $\{A_j\}_{j=1}^n$ in $\mathfrak{A}_0$ and scalars $\{\lambda_j\}$, the fact that U is an isometry implies

$$\| \sum_{j=1}^n \lambda_j K_{A_j} \|_q^p = \| \sum_1^n \lambda_j U \chi_{A_j} \|_q^p = \| U(\sum_1^n \lambda_j \chi_{A_j}) \|_q^p = \| \sum_1^n \lambda_j \chi_{A_j} \|_p^p.$$

Moreover,

$$\| \sum_{j=1}^n \lambda_j \chi_{A_j} \|_p^p = \sum_1^n |\lambda_j|^p \|\chi_{A_j}\|_p^p = \sum_1^n |\lambda_j|^p \|K_{A_j}\|_q^p,$$

and 3.4.2(iii) must hold. Finally,

$$\int_A H_B d\mu_1 = \overline{\langle \chi_A, U^* \chi_B \rangle} = \overline{\langle U \chi_A, \chi_B \rangle} = \int_B \overline{K_A} d\mu_2$$

which gives 3.4.2(iv) and completes the proof. $\qquad \square$

It is, perhaps, appropriate to observe that in most cases, there are no isometric embeddings of L^p into L^q when $p \neq q$, although it does happen for $1 \leq q < p \leq 2$. However, such an embedding cannot be positive. For, if $K_{A_1} \geq 0$, $K_{A_2} \geq 0$, then

$$|K_{A_1} + K_{A_2}|^q \geq |K_{A_1}|^q + |K_{A_2}|^q$$

and so

$$\|K_{A_1} + K_{A_2}\|_q^p \geq \left(\int |K_{A_1}|^q + \int |K_{A_2}|^q \right)^{\frac{p}{q}} = \left(\|K_{A_1}\|_q^q + \|K_{A_2}\|_q^q \right)^{\frac{p}{q}}$$
$$> \|K_{A_1}\|_q^p + \|K_{A_2}\|_q^p$$

because $\frac{p}{q} > 1$. This contradicts 3.4.2(iii).

3.5. Notes and Remarks

In his treatment of isometries on the L^p-spaces, Banach [21, pp.175-180] gives the details of the proof that isometries on $L^p[0,1]$ must take functions with disjoint support to functions with disjoint support; the argument for the discrete case (l^p, $1 \leq p < \infty$, $p \neq 2$) is dismissed as being analogous. He then states carefully the theorem which describes the isometries as being given by the canonical form (30). His description includes the form of the function h given there as having $|h|^p$ equal to what we would call a Radon-Nikodym derivative of $\lambda \cdot \varphi^{-1}$ where λ is Lebesgue measure. The proof is given for the l^p-spaces, and Banach remarks that the proof for $L^p[0,1]$ will appear in *Studia Mathematica IV*. As far as we know, this promised paper never appeared [21, p.178].

A general discussion of disjointness preserving linear operators on vector lattices and when they can be written in something like the canonical form (30) is given in the paper of Abramovich [1]. This paper includes several references to work of this type by Abramovich and his colleagues Kitover, Koldunov, Veksler, and Wickstead, among others. In particular, Abramovich considers the existence of the map φ so that $f(\varphi(s)) = 0$ implies $Tf(s) = 0$, which is so closely linked with disjointness preserving operators T and the canonical form. Lessard [196] also obtains results for disjointness preserving operators where the measure spaces are products of Souslin spaces. He gives a version of the Lamperti Theorem in that paper.

Lamperti's Results. Section 2 is an exposition of much of the oft-referenced paper by Lamperti [193], in which he undertakes to supply the missing proofs from Banach's work, and also to generalize the results to a broader class than $L^p[0,1]$. We have presented the general results of Lemma 3.2.1 and Theorem 3.2.2 from Lamperti's 1958 paper, but limited our application to the case of L^p. Lamperti also gives the following theorem.

3.5.1. THEOREM. *Suppose $\varphi(\sqrt{t})$ is either strictly convex or strictly concave and U is a linear operator on $L^\varphi(\Omega, \Sigma, \mu)$ such that $I_\varphi(Uf) = I_\varphi(f)$ for all $f \in L^\varphi$ (where I_φ is as defined in (34)). Then the conclusions of Theorem 3.2.5 hold. If in addition to the other hypotheses, φ is of regular variation at either $t = 0$ or $t = \infty$ but is not a power of t, then the set isomorphism T must be measure preserving, and $|h| = 1$ a.e. on $T(\Omega)$.*

The notion of regular variation at 0 and ∞ is due to Karamata [168]. To say that φ is of *regular variation* at $t = \infty$ means that $\lim\limits_{t \to \infty} \frac{\varphi(ct)}{\varphi(t)}$ exists for all $c > 0$ and similarly for $t = 0$. This implies that $\varphi(t) = t^p L(t)$ for some p where L has the property that $L(ct)/L(t) \to 1$ for $c > 0$. (We should mention that Lamperti [**193**] mixes up his references and thus refers to Karamata on page 400 when he means Hardy, Littlewood, and Polya [**127**].)

When $\varphi(t) = t^p$ in Theorem 3.2.2 we have what is usually called Clarkson's inequality [**74**]. As Lamperti points out, for $p \neq 2$ we cannot have an isometry from H^p onto L^p, since for functions in H^p, equality cannot hold in Clarkson's inequality unless $\|f\|_p = 0$ or $\|g\|_p = 0$. This answers a question which was posed by Boas [**40**].

It is natural to want to express an isometry in the more pleasing form of a composition operator as given by (30). This is possible when the regular set isomorphism T is given by a point mapping. The paper by Halmos and von Neumann [**123**] treats this question thoroughly. For example, T can be given by a point mapping when the measure space (Ω, Σ, μ) is complete and Ω is a metric space which is complete and separable, and with respect to which μ is a regular measure which is positive on open sets. Royden [**270**] also gives a good discussion of this question.

Perhaps this is a good place to mention a theorem of von Neumann [**314**].

3.5.2. THEOREM. *Every unitary operator U on $L^2[0,1]$ which satisfies $U(f)U(g) \in L^2$ and $U(fg) = U(f)U(g)$ whenever $f, g, fg \in L^2$ can be induced by an invertible measure preserving transformation on $[0,1]$.*

A proof of this statement is given by Choksi [**67**] who was interested in spectral properties of unitary operators on separable Hilbert spaces and when they can be represented as in von Neumann's Theorem. Choksi continues the discussion in a second paper [**68**] in which he drops the requirement that the transformation T be measure preserving. Here he gets results (and gives arguments) similar to Lamperti's. Of course there is a close relationship between the multiplicative property and the disjoint support condition.

Tulcea [**305**] observed in a footnote that a positive unitary operator on $L^2[0,1]$ is induced by an invertible measure preserving point transformation. A proof of this is given by Goodrich and Gustafson [**116**] who make use of von Neumann's Theorem. Choksi [**68**, p.797] remarks, in connection with this and his own Theorem 2, that it might be possible that for any unitary operator U on an abstract separable Hilbert space $\mathcal{H}$, there is a representation of $\mathcal{H}$ as L^2 so that U corresponds to an operator in canonical form.

The properties of the operator T_1 induced by the set isomorphism T are given by Doob in [**87**] as well as the explicit definition of T_1 as we have given it by (36). Doob actually assumes T is measure preserving but this is not necessary.

The proof of Lamperti's Theorem 3.2.5 is essentially the same as originally given, except we have treated the extension from the simple functions to all of L^p in a slightly different way. Our approach is suggested by the discussion in [**177**, p.697]. Also, we have allowed the range space to be a different L^p space, whereas Lamperti had assumed the operator to be from L^p to itself.

Lamperti's original statement of his theorem is slightly in error because of his equating of $|h|^p$ with the Radon-Nikodym derivative $\frac{d(\mu_1 \circ T^{-1})}{d\mu_2}$. (See [**57**], [**118**].) Example 3.2.6, which shows that this equality need not hold, is a modification to finite measure spaces of an example due originally to Grzaslewicz [**118**]. Lamperti had used the statement of his Theorem to obtain a corollary [**193**, Cor.3.1] which stated that if U is a transformation of measurable functions on (Ω, Σ, μ) which preserves the L^p norm for two values of p, then U is an isometry for all p. His argument used the canonical form (38) of the operator and the equation

$$|h|^{p_1} = \frac{d(\mu_1 \circ T^{-1})}{d\mu_2} = |h|^{p_2}$$

to conclude that $|h| = 1$ a.e. This need not hold as the example shows.

On the other hand, Example 3.2.6 involves an operator from one L^p space to another on a different underlying measure space, which is a slightly different setting than in Lamperti's work. The example of Grzaslewicz referred to above actually allows construction of an operator which is an isometry from $L^{p_1}(0, \infty) \cap L^{p_2}(0, \infty)$ into itself for $p_1 \neq p_2$ but which preserves the norm for no other p. Here is how it works.

Consider the regular set isomorphism T defined on the Lebesgue measurable subsets Σ of $(0, \infty)$ into the class of Lebesgue measurable subsets of $(-\infty, \infty)$ by

$$T(A) = -\alpha A \cup \beta A$$

where α, β are positive real numbers. For $0 < a < 1 < b$, let

$$h(t) = \begin{cases} a, & \text{for } t < 0; \\ b, & \text{for } t \geq 0. \end{cases}$$

For any r, $1 \leq r < \infty$, we have that

$$E\{|h|^r | T(\Sigma)\} = \frac{a^r \alpha + b^r \beta}{\alpha + \beta}.$$

For $a \leq p_1 < p_2 < \infty$, choose α, β, a, b as above so that

$$a^{p_1}\alpha + b^{p_1}\beta = a^{p_2}\alpha + b^{p_2}\beta = 1$$

and define U on $L^{p_1}(0, \infty) \cap L^{p_2}(0, \infty)$ into $L^{p_1}(-\infty, \infty) \cap L^{p_2}(-\infty, \infty)$ by

$$Uf(t) = \begin{cases} af(-t/\alpha) & \text{for } t < 0; \\ bf(t/\beta) & \text{for } t \geq 0. \end{cases}$$

Then U is an isometry for L^{p_1} and L^{p_2}, but not necessarily an isometry for any other $p \in [1, \infty)$. Since there exists an invertible measure preserving transformation from $(-\infty, \infty)$ onto $(0, \infty)$ (Grzaslewicz suggests $\tau(x) = |x + [x]|$ as one such transformation), it is possible to define an isometry from

$L^{p_1}(0,\infty) \cap L^{p_2}(0,\infty)$ into itself which shows that Lamperti's corollary fails. It would be interesting to find such an example for a finite measure space.

In one sense, the distinction between finite and infinite measure spaces is not so important. Indeed, if (Ω, Σ, μ) is σ-finite, and $1 \leq p < \infty$, then there is a probability measure μ_0 on (Ω, Σ) such that $L^p(\mu)$ and $L^p(\mu_0)$ are isometrically isomorphic. Furthermore, by a theorem of Caratheodory, there is an isometry from $L^p(\mu_0)$ to a subspace of $L^p([0,1], \Lambda, \lambda)$ which is onto if Σ has no atoms [62, p.28,40]. However, when we are interested in the actual form of isometries, the distinctions matter.

The paper of Grzaslewicz we have been discussing treats isometries on $L^1 \cap L^p$ with a special norm $|||f||| = max\{\|f\|_1, \|f\|_p\}$ and shows that an operator which is an isometry for both the L^1 and L^p norms extends to an isometry on all of L^1 and all of L^p. Hence it must have the canonical form. A later paper by Grzaslewicz and Schafer [119] treats the cases of $L^1 \cap L^\infty$ and $L^1 + L^\infty$ on $[0, \infty)$.

We wish to mention here some work by Carothers and several co-authors on the Lorentz spaces $L_{w,p}$ which are close relatives of the L^p-spaces. In [61] and [59] the authors use extreme point methods to characterize isometries on $L_{w,1}$. In [60] it is shown that an isometry on $L_{w,p}$ must satisfy the disjoint support property. Here is the main theorem from that paper.

3.5.3. THEOREM. *(Carothers, Hayden, and Lin) Let U be a linear isometry from $L_{w,p}$, $1 \leq p < \infty$, into itself. Then*

$$(Uf)^*(t) = sf^*(t/\alpha)$$

where $\alpha = \lambda(supp(U\chi_{(0,1)}))$ and $s = \frac{\|\chi_{(0,1)}\|}{\|\chi_{(0,\alpha)}\|}$. Moreover, if $s \neq 1$, then $w(\gamma) = s^p \alpha w(\alpha\gamma)$ almost everywhere.

In the statement above, λ represents Lebesgue measure, and f^* denotes the *decreasing rearrangement* of f which is involved in the definition of the Lorentz spaces. (See [200, pp.115ff].)

Generalizations of the Banach-Lamperti Theorem go in many directions. Russo [277] considered the isometries of $L^p(A, \phi)$ for $p = 1$ where ϕ is a faithful, finite, normal trace on a von Neumann algebra A and pointed out that Lamperti had settled the question for $1 < p < \infty$, $p \neq 2$ for the case where A is commutative and semifinite. Results of Yeadon [325] along these lines will be discussed in the second volume. Indeed, Lamperti's Theorem will appear in a variety of places in later chapters.

Subspaces of L^p and the Extension Theorem. Most of the proofs given in this section are taken directly from Hardin's paper [126]. Theorems 3.3.2, 3.3.3 and Corollary 3.3.7 are Hardin's versions of Theorems I and II in Rudin [275]. Throughout his paper, Hardin refers to the formulation given in 3.3.7 as "Rudin's Theorem." Rudin had considered only the complex case and had assumed M to be an algebra. Plotkin actually treated these questions earlier in a series of papers [255], [256], and [257] where he obtained similar results. Plotkin always assumed the subspace M to be contained in L^∞. He, too, treated the complex case, but he remarks at the end that his theorems remain valid in the real case with some slight modifications.

The difficulties caused by even values of p are interesting. The Example 3.3.8 was given by Rudin [275] who explains that it was suggested by a counterexample to the uniqueness of the Stieltjes moment problem [284], [304]. Plotkin [257] shows that his theorems can be proved for $p = 2k$ if the subspace M satisfies the following condition: there are at most k subalgebras $M_1, M_2, \ldots, M_k$ in L^∞ that contain the identity, are contained in M, and are such that their uniformly closed linear hull contains M. In particular, if M itself is an algebra contained in L^∞, then the condition is satisfied.

Other papers which are concerned with equimeasurability include those by Anderson [7], Schneider [280], Stephenson [292], and Al-Hussaini [4].

Although Plotkin [257] and Lusky [206] both treat the case of an extension of an isometry on a subspace of L^p to a larger subspace, Hardin [126] seems to have the most explicit description, both of the form of the isometry itself as given by Theorems 3.3.9, 3.3.14, and Corollary 3.3.15, and of the subspace to which the original isometry can be extended. The interested reader should also consult the work of Koldobsky and König [181] for a very nice, concise discussion of this problem, based mostly on Plotkin's papers. As an application of the extension theorem, we mention [180].

The notion of a function of full support and the fact that a subspace of L^p must have such a function (Lemma 3.3.11), was actually given by Ando [9, Lemma 3]. Hardin gave a different, independent argument and it is his which we have recorded in the proof of Lemma 3.3.11.

Hardin was motivated by some questions regarding the relationships between representations of symmetric stable stochastic processes and these ultimately became questions regarding isometries on subspaces spanned by the representing functions. Hardin notes that appropriate alterations in his proofs would make the theorems valid for the case where $\mathbb{F}$ is the field of quaternions.

Bochner Kernels. Plancherel's Theorem appears in a variety of places; the original appeared in 1910 [254]. Rudin [273, pp.187-189] gives a nice discussion of the theorem as an extension of the Fourier Transform from $L^1 \cap L^2$ to all of L^2 which is then an isometry from L^2 to itself. Bochner [41] mentions it as a special case of the formulas of Watson [317]. Watson had observed that his "kernel" operator represented a unitary if and only if a certain minimum condition holds, one like those given in the statement of Bochner's Theorem. But Watson's formulas did not include every unitary (e.g., the identity operator), and Bochner put together his own fomulation (Theorem 3.4.1) which does characterize all unitary transformations on L^2. We have followed the discussion of Bochner's Theorem given in Riesz and Nagy [266, pp.291-295].

The extension of Bochner's Theorem to the the case of operators from L^p to L^q was given by Merlo [**222**], and it was he who defined the *Bochner Kernels of type (p,q)*.

The question about whether one can embed a given L^p space inside of L^q is of considerable interest and many authors have treated it. We want to say enough here to justify our remark following the proof of Theorem 3.4.3. A good discussion is given in [**199**, pp.132-139], but for our purposes the clearest presentation is probably that of Banach himself.

3.5.4. THEOREM. [**21**, p.202] *If $L^p[0,1]$ is isomorphic to a subspace of $L^q[0,1]$, where $p > 1$, $q > 1$, then $q \leq p \leq 2$ or $2 \leq p \leq q$.*

From this we get immediately that

- If L^p is embeddable in L^q and conversely, then $p = q$.
- If $1 < p < 2 < q$ then neither of L^p or L^q is embeddable in the other.

Banach also proved that L^2 is embeddable in L^p for any p with $1 < p < \infty$. There are two cases left open:

(i) $q > p > 2$,
(ii) $1 \leq q < p \leq 2$.

The answer in the first case is negative, a result of Paley [**242**]. The second case has a positive answer, with the result being that for any such p and q, L^p is isometric to a subspace of L^q. For a proof (to which there were several contributors), see [**199**, p.139]. Other relevant references include [**188**], [**178**], [**197**], [**236**], and [**181**].

CHAPTER 4

Isometries of Spaces of Analytic Functions

4.1. Introduction

The form of an isometry of any Banach space is obviously determined by the geometry of the unit ball. The structure of the extreme points of the unit ball can be useful in characterizing the action of the surjective isometries of the space provided there aren't too many, e.g., as in the case of the unit ball of $L^p(\mu), 1 < p \neq 2 < \infty$, or too few, as in the case of c_0. In previous chapters we have seen how the extreme points of the *dual ball* can be used to obtain the form of isometries on the predual space. Another approach to the characterization of the surjective isometries of a Banach space X is to use the form of the extreme points of the unit ball of X along with the form of the isometries on the dual, X^*. Obviously, this requires knowledge of the isometries of the dual. We will illustrate this technique in the first section of this chapter by giving the deLeeuw-Rudin-Wermer solution of the surjective isometry problem for the space $H^1(\Delta)$, where Δ is the open unit disk in the complex plane. We follow this with the theorems of Forelli which give the form of all isometries of $H^p(\Delta)$ for $1 \leq p < \infty$. The following section will feature Kolaski's characterization of isometries on the Bergman spaces on the open unit disk.

The fourth section will be devoted to the Bloch spaces and we give the results of Cima and Wogen. In this setting the form of the extreme points of the unit ball of the Bloch space is not used. Furthermore, the isometries are not of *the canonical form* of a weighted composition operator.

In the final section we give an exposition of the work of Novinger and Oberlin on the isometries of the $S^p(\Delta)$. These spaces are very interesting from the point of view of isometries because the isometries are *integral operators*!

4.2. Isometries of the Hardy Spaces of the Disk

The first result on the isometries of the Hardy spaces was obtained for $H^\infty(\Delta)$ as a consequence of the characterization of the surjective isometries of function algebras on the disk. The proof for $H^1(\Delta)$ depends upon the structure of the extreme points of $H^1(\Delta)$ and the form of the isometries on $H^\infty(\Delta)$.

We begin by giving the result for the surjective isometries of $H^\infty(\Delta)$, which makes use of Theorem 2.3.16 of de Leeuw, Rudin, and Wermer on function algebras given in Chapter 2. We will also need a characterization of the automorphisms of $H^\infty(\Delta)$. By an automorphism we mean, of course, an algebra isomorphism.

4.2.1. LEMMA. *Suppose that T is an automorphism of $H^\infty(\Delta)$ onto itself. Then there is a conformal map τ of the disk onto itself such that $Tf(t) = f(\tau(t))$ for all $f \in H^\infty(\Delta)$. The converse also holds.*

PROOF. We first observe that for a function f in H^∞, a complex number λ is in the closure of the range of f if and only if $f - \lambda$ is not invertible in H^∞. Because T is an automorphism, it follows that $f - \lambda$ is invertible if and only if $Tf - \lambda$ is invertible so that the closure of the range of f and the closure of the range of Tf coincide. If q is the identity map on Δ, let $\tau = Tq$. Then the closure of the range of τ is the closed unit disk, and since τ is analytic, and therefore an open map, we must have $\tau(\Delta) \subset \Delta$.

Let $f \in H^\infty(\Delta)$ and $t \in \Delta$ be given. We must have

$$f - f(\tau(t)) = (q - \tau(t))g \text{ for some } g \in H^\infty,$$

and so

$$Tf - f(\tau(t)) = (\tau - \tau(t))Tg.$$

This last equation is zero at t, from which it follows that

$$Tf(t) = f(\tau(t)).$$

Since T^{-1} is also an automorphism of H^∞, from the previous argument, there is an analytic function ψ for which $T^{-1}g(t) = g(\psi(t))$ for all t. Clearly, ψ is the inverse of τ, and we conclude that τ is one-to-one and therefore conformal. In fact, τ must be a linear fractional transformation.   □

4.2.2. THEOREM. *(deLeeuw, Rudin, and Wermer) An operator T is a linear isometry of $H^\infty(\Delta)$ onto $H^\infty(\Delta)$ if and only if there is a unimodular complex number α and a conformal map τ of the disk such that*

$$Tf = \alpha(f \circ \tau)$$

for every $f \in H^\infty(\Delta)$.

PROOF. Since $H^\infty(\Delta)$ may be regarded as a subalgebra of $C(Q)$ for some compact, Hausdorff space Q, we may apply Theorem 2.3.16 and Lemma 4.2.1. The unimodular function h from the theorem is analytic on the disk and so must be a constant.   □

The extreme points of the unit ball of $H^1(\Delta)$ are given in the following theorem.

4.2.3. THEOREM. *(deLeeuw and Rudin) Let f be an H^1 function. Then f is an extreme point of the unit ball of H^1 if and only if f is an outer function of norm 1.*

Recall that an outer function f is a function of the form

$$f(z) = \lambda \exp\left[\frac{1}{2\pi}\int_{-\pi}^{\pi}\frac{e^{i\theta}+z}{e^{i\theta}-z}k(\theta)d\theta\right],$$

where $k(\theta)$ is a real L^1 function on the circle and λ is a complex number of modulus 1.

The following corollary is crucial in the proof of the isometry theorem of $H^1(\Delta)$.

4.2.4. COROLLARY. *(deLeeuw and Rudin) Let f be norm 1 function in H^1.*

(i) *If $\|f\|_1 = 1$ and f is not an extreme point of the unit ball then $f = \dfrac{1}{2}(f_1 + f_2)$, where f_1 and f_2 are distinct extreme points of the unit ball.*

(ii) *If $\|f\|_1 < 1$, then f is a convex combination of two extreme points of the unit ball.*

(iii) *The closure of the set of extreme points of the unit ball of H^1 consists of all H^1 functions f such that $\|f\|_1 = 1$ and f has no zeros in the disk.*

As we shall see later, all isometries of H^1 have been characterized by Forelli, but we give the proof of deLeeuw, Rudin, and Wermer for both historical and pedagogical reasons. The strategy is to first show that given the structure of the extreme points of H^1, every surjective isometry of H^1 must in fact be an isometry of H^∞ onto itself. This allows the application of the earlier results on isometries of function algebras.

4.2.5. THEOREM. *(deLeeuw, Rudin, and Wermer) An operator T is a linear isometry of H^1 onto H^1 if and only if T has the form*

$$(Tf)(z) = \alpha\tau^{'}(z)f(\tau(z))$$

where α is a complex number of modulus 1 and τ is a conformal map of the open unit disk onto itself.

PROOF. Since T is surjective, it takes extreme points to extreme points and thus takes the closure of the set of extreme points into itself. Any function f in H^1 which does not vanish in the disk is a multiple of an extreme point and consequently, Tf does not vanish in the disk. If we let $F(z) = T1(z)$ and if λ is a nonzero complex number in Δ, then the linearity of T implies that $f - \lambda$ has no zero in the disk if and only if $Tf - \lambda F$ does not vanish in the disk. Thus for every $f \in H^1$, we have $\{f(z) : z \in \Delta\} = \left\{\dfrac{Tf(z)}{F(z)} : z \in \Delta\right\}$. If we define $Uf(z) = (F(z))^{-1}Tf(z)$ for $f \in H^\infty$, then f and Uf have the same range. It is clear from the definition that $U : H^\infty \to H^\infty$ and U also preserves the H^∞ norm. Moreover, if g is any H^∞ function, the function $f = T^{-1}(Fg)$ belongs to H^∞ and $U(f) = g$. Therefore, U is a surjective isometry of H^∞ and $U(1) = 1$. This implies that U is an automorphism of H^∞ and hence, there exists a conformal map τ of the disk such that $U(f)(z) = f(\tau(z))$. This gives the following formula for T,

$$T(f)(z) = F(z)f(\tau(z)),$$

for every bounded function in H^1.

If $f \in H^\infty$,

$$\|f\|_1 = \frac{1}{2\pi} \int_{-\pi}^{\pi} |F(e^{i\theta})||f(\tau(e^{i\theta}))|d\theta.$$

For any $f \in H^1$, the norm is given by

$$\begin{aligned}\|f\|_1 &= \frac{1}{2\pi} \int_{-\pi}^{\pi} |f(e^{i\theta})|d\theta \\ &= \frac{1}{2\pi} \int_{-\pi}^{\pi} |\tau^{'}(e^{i\theta})||f(\tau(e^{i\theta}))|d\theta\end{aligned}$$

Given that T is an isometry, and that $g \circ \tau^{-1}$ belongs to H^∞ when $g \in H^\infty$ we obtain

$$\int_{-\pi}^{\pi} |F(e^{i\theta})||g(e^{i\theta})|d\theta = \int_{-\pi}^{\pi} |\tau^{'}(e^{i\theta})||g(e^{i\theta})|d\theta.$$

But for every bounded measurable function u on $|z| = 1$, there is a $g \in H^\infty$ such that $|g| = e^u$ for $|z| = 1$. Therefore the last equality implies that

$$\int_{-\pi}^{\pi} |F(e^{i\theta})|e^{u(\theta)}d\theta = \int_{-\pi}^{\pi} |\tau^{'}(e^{i\theta})|e^{u(\theta)}d\theta$$

for every bounded measurable function u on $|z| = 1$. Thus $|F(z)| = |\tau^{'}(z)|$ for $|z| = 1$. But both $F = T(1)$ and $\tau^{'}$ are outer functions and so we have $F(z) = \alpha\tau^{'}(z)$ where $|\alpha| = 1$. We have now shown that $Tf = \alpha\tau^{'}f \circ \tau$ for all $f \in H^\infty$. The density of H^∞ in H^1 yields the final step of the proof. □

As we have just seen, the extreme points are useful in determining the surjective isometries of H^1 but it is well known that every point of the unit ball of $H^p(\Delta)$ is an extreme point when $1 < p < \infty$. It is clear then that another approach is necessary to characterize the isometries of H^p when $1 < p < \infty$. It was Forelli who first discovered the key ideas for the solution of this problem. It is, perhaps, surprising but the key lemmas are results about measures and multiplicativity of linear maps.

In what follows we assume that σ_i, $i = 1, 2$, denote positive measures with total mass 1. We begin with a proposition whose generalization by Rudin was stated and proved in Section 3 of Chapter 3. We give another proof which is more directly adapted to the situation as we find it here.

4.2.6. PROPOSITION. *(Forelli, Rudin) Suppose that f_k is in $L^p(\sigma_k)$ ($k = 1, 2$) and that for all complex numbers z*

$$\int |1 + zf_1|^p d\sigma_1 = \int |1 + zf_2|^p d\sigma_2.$$

Then

$$\int |f_1|^2 d\sigma_1 = \int |f_2|^2 d\sigma_2$$

PROOF. Without loss of generality we may assume that

$$\int |f_1|^2 d\sigma_1 < \infty.$$

For $z \in \Delta$, set

$$g(z) = (1/2\pi) \int_0^{2\pi} |1 + ze^{ix}|^p dx - 1.$$

For $|z| < 1$ we can write,

$$|1 + ze^{ix}|^p = \sum_{j,k \geq 0}^{\infty} \binom{p/2}{j} \binom{p/2}{k} z^j \bar{z}^k e^{i(j-k)x}.$$

Thus,

$$g(z) = \binom{p/2}{1}^2 |z|^2 + \sum_{j=2}^{\infty} \binom{p/2}{j}^2 |z|^{2j}.$$

For $|z| \geq 1$ we can write,

$$g(z) = |z|^p \sum_{j=1}^{\infty} \binom{p/2}{j}^2 \left(\frac{1}{|z|}\right)^{2j}.$$

These equations imply that there are constants A_1, A_2 that depend only on p such that $g(z) \leq A_1 |z|^2$ for $|z| < 1$ while $g(z) \leq A_2 |z|^p$ for $|z| \geq 1$.

Furthermore, $\lim_{r \to 0} (1/r^2) g(r f_k) = \binom{p/2}{1}^2 |f_k|^2$.

From the inequalities above it is clear that

$$r^{-2} \left((1/2\pi) \left(\int_0^{2\pi} |1 + r f_k e^{ix}|^p dx \right) - 1 \right) \leq A_1 |f_k|^2 + A_2 |r|^{p-2} |f_k|^p$$

for $2 \leq p$. When $0 < p \leq 2$, the left hand side above is

$$r^{-2} \left((1/2\pi) \left(\int_0^{2\pi} |1 + r f_k e^{ix}|^p dx \right) - 1 \right) \leq A_3 |f_k|^2$$

for $0 < p \leq 2$, where A_3 also depends only on p.

Since $f_k \in L^p(\sigma_k)$ the dominated convergence theorem implies that with $k = 1$

$$\lim_{r \to 0} \int r^{-2} \left((1/2\pi) \int_0^{2\pi} |1 + r f_1 e^{ix}|^p dx - 1 \right) d\sigma_1 = (p^2/4) \int |f_1|^2 d\sigma_1$$

However, our hypothesis gives

$$\lim_{r \to 0} \int r^{-2} \left((1/2\pi) \int_0^{2\pi} |1 + rf_2 e^{ix}|^p dx - 1 \right) d\sigma_2$$

$$= \lim_{r \to 0} \int r^{-2} \left((1/2\pi) \int_0^{2\pi} |1 + rf_1 e^{ix}|^p dx - 1 \right) d\sigma_1.$$

Since $g(rf_k) \geq 0$ and $r^{-2} g(rf_k) \to (p^2/4)|f_k|^2$, Fatou's lemma implies (together with the previous statement) that

$$(p^2/4) \int |f_2|^2 d\sigma_2$$

$$\leq \liminf_{r \to 0} r^{-2} \left((1/2\pi) \int \left(\int_0^{2\pi} |1 + rf_2 e^{ix}|^p dx - 1 \right) d\sigma_2 \right)$$

$$= \liminf_{r \to 0} r^{-2} \left((1/2\pi) \int \left(\int_0^{2\pi} |1 + rf_1 e^{ix}|^p dx - 1 \right) d\sigma_1 \right)$$

$$= (p^2/4) \int |f_1|^2 d\sigma_1.$$

Therefore we have shown that

$$\int |f_2|^2 d\sigma_2 \leq \int |f_1|^2 d\sigma_1$$

and so $\int |f_2|^2 d\sigma_2 < \infty$ as well. Repeat the argument with the roles of f_2 and f_1 interchanged and we obtain $\int |f_1|^2 d\sigma_1 \leq \int |f_2|^2 d\sigma_2$ and the proof is complete.

□

4.2.7. PROPOSITION. *(Forelli) Let $\mathcal{A}$ be a subalgebra of $L^\infty(\sigma_1)$ that contains the constants, and let T be a linear transformation of $\mathcal{A}$ into $L^\infty(\sigma_2)$ with $T1 = 1$. Suppose that $p \neq 2$ and*

$$\int |Tf|^p d\sigma_2 = \int |f|^p d\sigma_1$$

for all $f \in \mathcal{A}$. Then T is multiplicative:

$$T(fg) = TfTg.$$

PROOF. To prove that T is multiplicative on $\mathcal{A}$, it is sufficient to prove that $T(f^2) = (Tf)^2$ whenever $f \in \mathcal{A}$. To that end consider the following,

$$\int |(Tf)^2 - T(f^2)|^2 d\sigma_2$$

$$= \int [(Tf)^2 + T(f^2)][\overline{(Tf)^2} - \overline{T(f^2)}] d\sigma_2$$

$$= \int |Tf|^4 d\sigma_2 - 2\mathrm{Re} \int (Tf)^2 \overline{T(f^2)} d\sigma_2 + \int |T(f^2)|^2 d\sigma_2.$$

We will show that the right hand side of this last equation is zero by showing that each one of the integrals on the right is equal to $\int |f|^4 d\sigma_1$.

For $f \in \mathcal{A}$, $T(1 + zf) = 1 + zTf$ and

$$\int |1 + zTf|^p d\sigma_2 = \int |1 + zf|^p d\sigma_1.$$

For $|z|$ small we can write,

$$|1 + zf|^p = \sum_{j,k \geq 0} \binom{p/2}{j} \binom{p/2}{k} z^j \bar{z}^k f^j \bar{f}^k$$

$$\int |1 + zf|^p d\sigma_1 = \sum_{j,k \geq 0} \binom{p/2}{j} \binom{p/2}{k} z^j \bar{z}^k \int f^j \bar{f}^k d\sigma_1.$$

Writing the corresponding equation for $\int |1 + zTf|^p d\sigma_2$ and using the hypothesis yields

$$\sum_{j,k \geq 0} \binom{p/2}{j} \binom{p/2}{k} \int (Tf)^j \overline{(Tf)}^k d\sigma_2 = \sum_{j,k \geq 0} \binom{p/2}{j} \binom{p/2}{k} \int f^j \bar{f}^k d\sigma_1.$$

Therefore,

$$\binom{p/2}{j} \binom{p/2}{kz^j} \bar{z}^k \int (Tf)^j \overline{(Tf)}^k d\sigma_2 = \binom{p/2}{j} \binom{p/2}{kz^j} \bar{z}^k \int f^j \bar{f}^k d\sigma_1.$$

Since $p \neq 2$ we can conclude for $j = k = 1$ that

$$\int (Tf) \overline{Tf} d\sigma_2 = \int f \bar{f} d\sigma_1 \text{ or}$$

$$\int |Tf|^2 d\sigma_2 = \int |f|^2 d\sigma_1.$$

Replace f by f^2 in this last equation and we obtain that

$$\int |T(f^2)|^2 d\sigma_2 = \int |f|^4 d\sigma_1.$$

When $j = k = 2$ we obtain that

$$\int |Tf|^4 d\sigma_2 = \int |f|^4 d\sigma_1.$$

For $j = 2, k = 1$ we get

$$\int (Tf)^2 \overline{Tf} d\sigma_2 = \int f^2 \bar{f} d\sigma_1.$$

Replace f by $f + zg$ in this last equation and equate the coefficients of the resulting polynomial in z to obtain

$$\int (Tf)^2 \overline{Tg} d\sigma_2 = \int f^2 \bar{g} d\sigma_1.$$

This last equation holds for all f and g in $\mathcal{A}$ and in particular for $g = f^2$. Hence,

$$\int (Tf)^2 \overline{T(f^2)} d\sigma_2 = \int f^2 \bar{f}^2 d\sigma_1$$

$$= \int |f|^4 d\sigma_1.$$

Therefore,

$$\int |T(f^2) - (Tf)^2|^2 d\sigma_2 = 0$$

for every $f \in \mathcal{A}$ and the proof is complete.

$\square$

As we have seen, it is often more difficult to obtain results about injective isometries than the surjective ones. For example in the case of the $H^\infty(\Delta)$, only the form of the surjective isometries are known. However, for $H^p(\Delta)$ Forelli was able to characterize both classes when $1 \leq p < \infty, p \neq 2$.

Before stating this next result we pause for a little notation. We denote by Σ the class of all σ-measurable sets of the circle, where σ is normalized Lebesgue measure on the circle. If ϕ is any nonconstant inner function we define,

$$\Sigma(\phi) = \{X \Delta Z : X \in \phi^{-1}(\Sigma) \text{ and } Z \in \Sigma, \sigma(Z) = 0\}.$$

We denote by $\mu \equiv \sigma \circ \phi^{-1}$ the measure induced on the Borel sets of the unit circle by ϕ. For every Borel measurable function g on the unit circle we have

$$\int g d\mu = \int g(\phi) d\sigma.$$

The Fourier-Stieltjes coefficients of this measure are given by

$$\int \overline{\phi^n} d\sigma.$$

Since the map $g \to \int g d\sigma$ is a multiplicative linear functional on H^∞,

$$\int \phi^n d\sigma = \left(\int \phi d\sigma \right)^n$$

when $n > 0$. Hence the Fourier-Stieltjes coefficient of μ at a positive n is

$$\left(\int \overline{\phi} d\sigma \right)^n = \int e^{int} P(e^{it}) d\sigma(t),$$

where $P(z) = \dfrac{(1 - |a|^2)}{|1 - az|^2}$ is the Poisson kernel with $a = \int \overline{\phi} d\sigma$

4.2.8. THEOREM. *(Forelli) Suppose that $p \neq 2$ and that T is a linear isometry of H^p into H^p. Then there is a nonconstant inner function ϕ and an $F \in H^p$ such that*

$$T f = F f \circ \phi$$

for every $f \in H^p$. The functions ϕ and F are related by the relation

$$\int_X |F|^p d\sigma = \int_X \frac{d\sigma}{P(\phi)}$$

whenever $X \in \Sigma(\phi)$, where P is the Poisson kernel induced by ϕ. Conversely, when a nonconstant inner function ϕ and a function $F \in H^p$ are related by this relation, $T f = F f \circ \phi$ defines a linear isometry of H^p into itself.

PROOF. Let T be an isometry of H^p into H^p. Set $F = T(1)$, and define a measure ν on the circle by $d\nu = |F|^p d\sigma$. Since F is an H^p function, it can't vanish on any set of σ-positive measure. Consequently, ν and σ are mutually absolutely continuous measures on the circle.

Define $S : H^p \to L^p(\nu)$ by $Sf = \dfrac{Tf}{F}$. Then S is an isometry of H^p into $L^p(\nu)$ with $S(1) = 1$. Let $\chi(z) = z$. Then,

$$\int |S(\chi^n)|^p d\nu = 1$$

and by Proposition 4.2.6,

$$\int |S(\chi^n)|^2 = 1.$$

It follows from equality in Hölder's inequality that $|S(\chi^n)| = 1$ a.e. Thus S maps the algebra generated by χ into $L^\infty(\nu)$. From the Proposition 4.2.7 above we know that S is multiplicative on this algebra and so if p is any polynomial, $S(p(\chi)) = p(S(\chi))$ and

$$T(p(\chi)) = F p(\phi)$$

where $\phi = S(\chi)$.

Since $F \in H^p$, $F = MG$, where M is an inner function and G is an outer function belonging to H^p. It follows from the form of T that $M\phi^n$ is an inner function when $n \geq 0$. Moreover, ϕ is also an inner function. To see this let

B denote the closed subspace of $L^2(\sigma)$ spannned by $\chi^j \phi^k$ $(j, k \geq 0)$. Clearly **B** is invariant under multiplication by χ. Since $M\mathbf{B} \subseteq H^2$, **B** is not closed under multiplication by $\overline{\chi}$. It follows from a result given in Hoffman [**132**, p. 102] that $\mathbf{B} = \psi H^2$, where $|\psi| = 1$. Thus, $\chi^j \phi^k \psi$ belongs to **B** for $j, k \geq 0$. Since $1 \in \mathbf{B}$, ψ is in the closure of the polynomials in χ and ϕ. Therefore, ψ^2 belongs to **B** and hence $\psi^2 = \psi g$ where $g \in H^2$. It follows that ψ belongs to H^2 and it must be that $\mathbf{B} = H^2$ which implies that ϕ is inner.

We have that $Tp = Fp(\phi)$ for all polynomials and that ϕ is a nonconstant inner function. Since T is bounded, the density of these polynomials in H^p imply that $Tf = Ff(\phi)$ for all $f \in H^p$. For every f in H^p we have

$$\int |F|^p |f(\phi)|^p d\sigma = \int |f|^p d\sigma.$$

If we let $X = \phi^{-1}(Y)$ where $Y \in \Sigma$, this last equation implies that

$$\int_X |F|^p d\sigma = \int_Y d\sigma$$
$$= \int_Y (d\sigma/d\mu)\, d\mu$$
$$= \int_X \frac{1}{P(\phi)} d\sigma.$$

Since the converse is clear, the proof of the theorem is finished.   □

For onto isometries, the description of T is a bit sharper.

4.2.9. THEOREM. *(Forelli) Suppose that $p \neq 2$ and T is a linear isometry of H^p onto H^p. Then*

$$Tf = b(d\phi/dz)^{1/p} f(\phi)$$

where ϕ is a conformal map of the unit disk onto itself and b is a unimodular complex number. Conversely, if T, b, and ϕ are given as above then T is a surjective isometry.

PROOF. Since T and T^{-1} are both isometries, we know that there exist nonconstant inner functions ϕ and ψ together with H^p functions F and G such that $Tf = Ff(\phi)$ and $T^{-1}f = Gf(\psi)$. Since $TT^{-1}f = T^{-1}Tf = f$ we have

$$FG(\phi)f(\psi) = GF(\psi)f(\psi) = f.$$

When $f = 1$, we see that $FG(\phi) = GF(\psi)$ and hence $f(\psi(\phi)) = f(\phi(\psi)) = f$ for every f belonging to H^p. This implies that ϕ is a conformal map of the disk onto itself and consequently $\Sigma = \Sigma(\phi)$. Furthermore, $|d\phi/dz| = \dfrac{1}{P(\phi)}$ where P is the Poisson kernel induced by ϕ. The isometry condition requires that

$$\int_X |F|^p d\sigma = \int_X |d\phi/dz| d\sigma$$

when X belongs to Σ. Hence $|F| = |(d\phi/dz)^{1/p}|$. This equality in addition to the fact that $(d\phi/dz)^{1/p}$ is an outer function makes it obvious that F is also outer. Therefore, $F = \lambda(d\phi/dz)^{1/p}$ where λ is a complex number of modulus 1 and the proof is complete. □

4.3. Bergman Spaces

Another class of analytic function spaces that has been of interest for some time is the class of Bergman spaces. We will restrict our focus to the Bergman spaces on the open unit disk Δ. For $0 < p < \infty$, the Bergman p-space L_a^p on the unit disk Δ is the collection of all functions analytic on Δ for which

$$\int |f(z)|^p d\mu(z) < \infty,$$

where μ is normalized Lebesgue measure on the unit disk. We will see that the isometries on these spaces are again described as weighted composition operators.

4.3.1. THEOREM. *(Kolaski) Suppose T is a linear isometry from $L_a^p(\Delta)$ to itself. Then there exist analytic functions $\varphi : \Delta \to \Delta$ and $h \in L_a^p$ such that*

$$(51) \qquad Tf(z) = h(z)(f \circ \varphi)(z)$$

for all $z \in \Delta$ and $f \in L_a^p$. Furthermore, $\varphi(\Delta)$ is dense in Δ and

$$(52) \qquad \int (g \circ \varphi)|h|^p d\mu = \int g d\mu$$

for every bounded Borel function g on Δ. If T is surjective, then φ is an automorphism of Δ and h is related to φ by

$$(53) \qquad |h^p| = |J_\varphi|$$

where J_φ is the Jacobian of φ.

Conversely, if the conditions above are satisfied, then T is an isometry which is surjective if (53) holds.

PROOF. Let $h = T1$ and define $\nu = |h|^p d\mu$. For $f \in L_a^\infty$, define

$$Af = \frac{Tf}{h}.$$

Since h is not zero in L^p_a, it is necessarily nonzero a.e. (μ), and so Af makes sense. If $f \in L^\infty_a$ with $\|f\|_\infty \le 1$, then

$$esssup_{z \in \Delta} |Af(z)| = \lim_{n \to \infty} \left(\int |Af|^{pn} |h|^p d\mu(z) \right)^{\frac{1}{pn}}$$

$$= \lim_{n \to \infty} \left(\left\{ \int |A(f^n)(z)h(z)|^p d\mu \right\}^{\frac{1}{p}} \right)^{\frac{1}{n}}$$

$$= \lim \left(\left\{ \int |T(f^n)|^p d\nu \right\} \frac{1}{p} \right)^{\frac{1}{n}}$$

$$\le \limsup_{n \to \infty} \left(\left\{ \int |f^n|^p d\nu \right\}^{\frac{1}{p}} \right)^{\frac{1}{n}} \le 1.$$

Hence, for any $f \in L^\infty_a(\Delta)$, the function Tf/h is an essentially bounded meromorphic function and consequently is analytic. It follows as in Proposition 4.2.7 that A is a multiplicative linear map of L^∞_a into $L^\infty(\nu) = L^\infty(\mu)$ which preserves the supremum norms. Hence, A is a multiplicative linear sup norm isometry of L^∞_a into itself with $A1 = 1$.

Let q denote the identity function on Δ and define φ on Δ by $\varphi(z) = Aq(z)$. Note that for any $z \in \Delta$,

$$\int |1 + z\varphi|^p d\nu = \int |A(1 + zq)|^p d\nu = \int |1 + zq|^p d\mu.$$

Arguments like those in the proof of Proposition 4.2.7 will show that if g is any polynomial in the variables $z, \bar{z}$, then

(54)
$$\int_\Delta (g \circ \varphi) d\nu = \int_\Delta g d\mu.$$

It now follows from standard integral convergence theorems that (54) holds for any continuous function and then also for the characteristic function of any Borel set in $\mathbb{C}$. This means that φ and q are equimeasurable, so that (52) holds for every bounded Borel function g.

We want to show now that the range of the analytic function φ is dense in Δ. Let $\Omega = \varphi^{-1}(\Delta)$. Our calculations above show that

(55)
$$\nu(\varphi^{-1}(E)) = \mu(E)$$

for every Borel set in $\mathbb{C}$, so that

$$1 = \mu(\Delta) = \nu(\Omega).$$

Recall that

$$\nu(\Delta) = \int d\nu = \int |h|^p d\mu = \|T1\| = 1.$$

It is straightforward to show that μ is absolutely continuous with respect to ν, so that $\nu(\Delta \backslash \Omega) = 0$ implies that $\mu(\Delta \backslash \Omega) = 0$. Since φ is continuous, Ω is open, and therefore dense in Δ. (If Ω were not dense, there would be an open disk inside $\Delta \backslash \Omega$, contrary to the fact that it has Lebesgue measure zero.) This density implies that $\varphi(\Delta) \subset \overline{\varphi(\Omega)} \subset \overline{\Delta}$. Putting $\varphi(\Delta)$ for E in (55) shows that

$$\mu(\varphi(\Delta)) = \nu(\Delta) = 1.$$

This, together with the fact that $\varphi(\Delta) \subset \overline{\Delta}$, means that $\varphi(\Delta)$ is dense in Δ.

By the definition and multiplicativity of A, we see that

$$TP(z) = h(z)(P \circ \varphi)(z)$$

for all polynomials P and $z \in \Omega$. Since the polynomials are dense in L_a^p and the point evaluations $\hat{z}(f) = f(z)$ are continuous linear functionals on L_a^p, (as seen by means of the Cauchy integral theorem), for each f in the Bergman space, there is a sequence $\{P_n\}$ of polynomials such that $P_n(z) \to f(z)$. From this, and the statement for polynomials above, we obtain (51) for all $z \in \Omega$.

For the first part of the theorem to be proved, it remains to show that $\Omega = \Delta$. Let $z \in \Delta$ with $h(z) \neq 0$. By density of Ω, there is a sequence $\{z_n\}$ in Ω such that $z_n \to z$, and therefore,

$$(f \circ \varphi)(z_n) = \frac{Tf(z_n)}{h(z)} \to \frac{Tf(z)}{h(z)}.$$

Thus (51) holds for such z, and $\varphi(z) \in \Delta$ since Δ is a domain of analyticity. Now suppose there is some z_0 with $h(z_0) = 0$ and $|\varphi(z_0)| = 1$. Choose a small disk $\overline{D}$ which contains no zero of h except z_0. Then $\varphi(\partial D)$ is a compact subset of Δ, and there exists a polynomial P satisfying

$$|P(\varphi(z_0))| > \sup_{t \in \varphi(\partial D)} \{|P(t)|\}.$$

Then $T(P^n) = h(P^n \circ \varphi)$ will not obey the maximum principle on $\overline{D}$ near z_0. This contradiction completes the proof that $\Omega = \Delta$.

If we suppose that T is onto, then what we have previously proved applies to T^{-1} so that

$$T^{-1}f = g(f \circ \psi)$$

where $g = T^{-1}1$. For any $f \in L_a^p$ we must have

$$f = TT^{-1}f = h(g \circ \varphi)(f \circ \psi \circ \varphi)$$

and

$$f = T^{-1}Tf = g(h \circ \psi)(f \circ \varphi \circ \psi).$$

Putting $f = 1$ in each equation, and then $f = q$, we conclude that $h(g \circ \varphi) = 1 = h(g \circ \psi)$ and $\psi = \varphi^{-1}$. Hence, φ is conformal and and an automorphism of Δ.

Finally, we let f be continuous and by putting $g = f \circ \varphi^{-1}$, we obtain

$$\int_\Delta f|h|^p d\mu = \int_\Delta f \circ \varphi^{-1} d\mu = \int_\Delta f|J_\varphi| d\mu$$

by the change of variables property in the integral. Since the above holds for all continuous functions f, we find $|h|^p = |J_\varphi|$ a.e. (μ), and (53) follows by the continuity of h and J_φ.

The proof of the converse statements is straightforward.

$\square$

It is interesting to ask about the nature of the analytic functions φ which can give isometries of the Bergman space as described above. For example, φ may not map Δ onto all of Δ. If we let

(56) $$\varphi(z) = -1 + 2(1 + iz)/\psi(z)$$

where

(57) $$\psi(z) = 1 + iz + \sqrt{2(1 - z^2)}.$$

For this definition of ψ we use the principal branch for the square root. Then φ maps Δ in a one-to-one fashion onto $\Delta\backslash[0,1)$. If we let $h(z) = [d\varphi/dz]^2$, then $h \in L_a^1(\Delta)$ and since $|h| = |J_\varphi|$, we have that φ and h satisfy (52), and so determine an isometry on $L_a^1(\Delta)$ by means of (51).

The function $\varphi(z) = z^2$ can be shown to generate an isometry of $L_a^1(\Delta)$, so that φ need not be one-to-one. However, it is also known that $\varphi(z) = z^2$ cannot generate isometries of $L_a^p(\Delta)$ for $p \neq 1, 2$. See the notes at the end of the chapter for more information on this.

4.4. Bloch Spaces

The isometries of the Hardy spaces and the Bergman spaces on the disk have the canonical form of a weighted composition operator. We now introduce a class of Banach spaces of analytic functions, called the Bloch spaces, for which the isometries are not quite in the canonical form. The geometry of the unit ball and the form of the isometries on Bloch spaces were first studied by Cima and Wogen, and we follow their development in this section.

The Bloch space is defined as follows: Let Δ denote the open unit disc in the complex plane and let Γ denote the unit circle. The set $\mathcal{B}$ of Bloch functions is the set of holomorphic functions on Δ defined by

$$\mathcal{B} = \{f : f(0) = 0, \ sup\{|f'(z)|(1 - |z|^2) : z \in \Delta\} < \infty\}.$$

The space $\mathcal{B}$ with norm defined by,

$$\|f\| = \sup\{|f(z)|(1 - |z|^2) : z \in \Delta\}$$

is a nonseparable Banach space. We let $\mathcal{B}_0$ denote the subspace of $\mathcal{B}$ spanned by the polynomials. The structure of the isometries on $\mathcal{B}_0$ is quite surprising. We will show that every isometry of this Banach space must be surjective.

Before we proceed with the proof we want to introduce some notation for some particular subsets of the continuous functions on the open disk Δ. First, we let

$$\mathcal{C} = \{f \in C(\Delta) : \|f\|_\mathcal{C} \equiv sup\{|f(z)|(1 - |z|^2) : z \in \Delta\} < \infty\}.$$

If $\mathcal{D}$ denotes the functions in $\mathcal{C}$ that are holomorphic on Δ, then $\mathcal{D}$ is a closed subspace of $\mathcal{C}$ and the derivative operator $D(f) = f'$ is a linear isometry of $\mathcal{B}$ onto $\mathcal{D}$. Set $\mathcal{D}_0 = D(\mathcal{B}_0)$.

Define a mapping $\Phi : \mathcal{C}_b(\Delta) \to \mathcal{C}$ by

$$(\Phi f)(z) = f(z)(1 - |z|^2)^{-1}.$$

It is easily seen that Φ is a surjective linear isometry and we denote Φ^{-1} by Ψ. Let $\mathcal{A} = \Phi(\mathcal{D})$ and $\mathcal{A}_0 = \Phi(\mathcal{D}_0)$. It follows that

$$\mathcal{A} = \{f'(z)(1 - |z|^2) : f \in B\}, \quad \mathcal{A}_0 = \{f'(z)(1 - |z|^2) : f \in \mathcal{B}_0\}.$$

We also have the following set inclusions:

$$\mathcal{A}_0 \subseteq \mathcal{A} \subseteq \mathcal{C}_b(\Delta) \quad \text{and} \quad \mathcal{D}_0 \subseteq \mathcal{D} \subseteq C(\Delta).$$

Finally, note that

$$\mathcal{D}_0 \subseteq C_0(\Delta),$$

where $C_0(\Delta)$ is all continuous functions on Δ which vanish on Γ.

4.4.1. PROPOSITION. *Suppose that* $\tau : \Delta \to \Delta$ *is analytic. Then there is an analytic function* f *on* Δ *with*

$$|f(z)| = (1 - |\tau(z)|^2)(1 - |z|^2)^{-1}$$

if and only if $\tau(z) = \lambda(z - \alpha)(1 - \alpha z)^{-1}$.

PROOF. If τ is a conformal map of Δ, then by the Schwartz-Pick lemma, $f(z) = \tau'(z)$ satisfies the condition. Now, if such a function f satisfies the condition above, then the function $\log |f(z)| = \log (1 - |\tau(z)|^2) - \log (1 - |z|^2)$ is harmonic on Δ. A computation shows that

$$\triangle \log (1 - |\tau(z)|^2) = \triangle \log (1 - |z|^2)$$

reduces to

$$\frac{|\tau'(z)|^2}{(1 - |\tau(z)|^2)^2} = \frac{1}{(1 - |z|^2)^2}.$$

Thus,

$$|\tau'(z)| = (1 - |\tau(z)|^2)(1 - |z|^2)^{-1}$$

and again by the Schwartz-Pick lemma, τ must be of the desired form. $\square$

4.4.2. PROPOSITION. *Every* $\alpha \in \Delta$ *is a peak point of* $\mathcal{A}_0$.

PROOF. Let $f_\alpha(z) = (1 - |z|^2)(1 - \bar{\alpha}z)^{-2}$. The function f_α belongs to $\mathcal{A}_0$ and f_α peaks at α. $\square$

We recall here that the set of extreme points of the unit ball of a Banach space X is denoted by $ext(X)$.

4.4.3. THEOREM. *Let* $S : \mathcal{D}_0 \to \mathcal{D}_0$ *be an isometry. Then there is a conformal automorphism* ϕ *of* Δ *and a* $\lambda \in \Gamma$ *so that* $Sf(z) = (\lambda\phi')(z)(f(\phi(z)))$

PROOF. We prove the sufficiency first. To that end, let $g \in \mathcal{D}_0$.

$$\|Sg\|_{\mathcal{D}_0} = sup\{|S(g)(z)|(1-|z|^2) : z \in \Delta\}$$
$$= sup\{|\phi'(z)||g(\phi(z))| : z \in \Delta\}.$$

By Proposition 4.4.1 this last equation can be written as

$$\|S(g)\|_{\mathcal{D}_0} = sup\left\{|\phi'(z)g(\phi(z))|\frac{1-|\phi(z)|^2}{|\phi(z)|} : z \in \Delta\right\}$$
$$= sup\{g(z)(1-|z|^2) : z \in \Delta\}$$
$$= \|g\|_{\mathcal{D}_0}.$$

For the necessity of these conditions, we suppose that $S : \mathcal{D}_0 \to \mathcal{D}_0$ is an isometry. Recall that $\mathcal{A}_0 = \{f'(z)(1-|z|^2) : f \in \mathcal{B}_0\}$ and that $\Psi = \Phi^{-1}$. Define $T = \Psi S\Phi|_{\mathcal{A}_0}$. Then T is an isometry form $\mathcal{A}_0$ onto its range which we denote by $\mathcal{R}_0$. The adjoint T^*, $T^* : \mathcal{R}_0^* \to \mathcal{A}_0^*$ is a surjective isometry and consequently $T^* : ext(\mathcal{R}_0)^* \to ext(\mathcal{A}_0)^*$. The standard argument along with Proposition 4.4.2 gives us that the extreme points are of the form $\lambda\psi_z$, where ψ_z is the evaluation function at z.

Let $\Sigma(\mathcal{R}_0) = \{z \in \Delta : \psi_z|_{\mathcal{R}_0} \in E^\Phi(\mathcal{R}_0^*)$. Hence, there are functions

$$\tau : \Sigma(\mathcal{R}_0) \to \Delta \quad \text{and} \quad \alpha : \Sigma(\mathcal{R}_0) \to \Gamma$$

so that

$$T^*(\psi_z|_{\mathcal{R}_0}) = \alpha(z)(\psi_{\tau(z)}|_{\mathcal{A}_0}) \quad \text{for all} \quad z \in \Sigma(\mathcal{R}_0).$$

Therefore, for all $z \in \Sigma(\mathcal{R}_0)$ and $f \in \mathcal{A}_0$, T is of the form $Tf(z) = \alpha(z)f(\tau(z))$. In particular, for $k \in \mathbf{Z}^+$ and $z \in \Sigma(\mathcal{R}_0)$,

$$T(z^k(1-|z|^2)) = \alpha(z)(\tau(z))^k(1-|\tau(z)|^2).$$

Hence, we can define a sequence of functions $G_k \in \mathcal{D}_0$ holomorphic on Δ, such that

$$T(z^k(1-|z|^2)) \equiv G_k(z)(1-|z|^2).$$

Now,

$$G_0(z) = \frac{\alpha(z)(1-|\tau(z)|^2)}{1-|z|^2} \quad \text{and} \quad G_1(z) = \frac{\alpha(z)\tau(z)(1-|\tau(z)|^2)}{1-|z|^2}.$$

Clearly, $\tau(z) = \frac{G_0(z)}{G_1(z)}$ and hence τ has a meremorphic extension from $\Sigma(\mathcal{R}_0)$ to all of Δ. This extension is unique since $\Sigma(\mathcal{R}_0)$ is uncountable.

If we define $p_n(z) = z^n$, then $p_n \in \mathcal{B}_0$ and it follows that for all $z \in \Sigma(\mathcal{R}_0)$,

$$Sp_n = G_0(z)\left(\frac{G_1(z)}{G_0(z)}\right)^n$$

Since the extension from $\Sigma(\mathcal{R}_0)$ is unique, this equation holds for all $z \in \Delta$.

It follows that $\dfrac{G_1}{G_0}$ has no poles and thus the meromorphic extension of τ is actually holomorphic. Moreover,

$$\|p_n\|_{\mathcal{D}} = \left(\frac{n}{n+2}\right)^{n/2}\left(\frac{2}{n+2}\right) < 1;$$

hence, $\|p_n\|_{\mathcal{D}} < 1$ for $n \geq 0$. Thus, $|G_0(z)||\dfrac{G_1(z)}{G_0(z)}|^n(1-|z|^2) \leq \|Sp_n\|_{\mathcal{D}} < 1$ for all $z \in \Delta$ and $n \geq 0$. Therefore, the range of $\dfrac{G_1}{G_0}$ is contained in Δ. From Proposition 4.4.1 we deduce the existence of a conformal automorphism of Δ such that $G_0 = \lambda\phi'$ for some $\lambda \in \Gamma$. Therefore, for each polynomial p we have $Sp(z) = (\lambda\phi'(z))p(\phi(z))$ and the proof is completed by appealing to the density of the polynomials in $\mathcal{B}_0$. $\qquad\square$

4.4.4. COROLLARY. *If $S : \mathcal{B}_0 \to \mathcal{B}_0$ is an isometry, then there is a conformal automorphism ϕ of Δ and a $\lambda \in \Gamma$ so that $Sf(z) = \lambda(f(\phi(z)) - f(\phi(0))$ for all $f \in \mathcal{B}_0$.*

PROOF.

$$
\begin{aligned}
DSD^{-1}f'(z) &= \lambda\phi'(z)f'(\phi(z)) \\
&= \lambda(f(\phi))'(z). \\
D(Sf)(z) &= \lambda(f \circ \phi)'. \\
Sf(z) &= \int_0^z \lambda(f \circ \phi)'(\xi)d\xi. \\
Sf(z) &= \lambda[f \circ \phi(z) - f \circ \phi(0)].
\end{aligned}
$$

$\qquad\square$

4.4.5. COROLLARY. *Every isometry of $\mathcal{B}_0$ is surjective.*

Cima and Wogen also give characterizations for the surjective isometries of the big Bloch space $\mathcal{B}$. We will only state their results in this setting and refer the interested reader to their paper for the complete details. The proof is similar to the one for the little Bloch space, but requires the use of the Stone-Čech compactification of Δ.

4.4.6. THEOREM. *Let $S : \mathcal{D} \to \mathcal{D}$ be a surjective isometry. Then there is a conformal automorphism ϕ of Δ and $\lambda \in \Gamma$ such that $Sf = (\alpha\phi')(f \circ \phi)$.*

They also obtain the following corollary.

4.4.7. COROLLARY. *If $S : \mathcal{B} \to \mathcal{B}$ is a surjective isometry, then there is a conformal automorphism ϕ of Δ and a $\lambda \in \Gamma$ such that*

$$Sf = \lambda(f \circ \phi - f(\phi(0))).$$

4.5. S^p Spaces

As we have seen, the Bloch spaces admit isometries which are not weighted composition operators. We are going to describe another class of Banach spaces which admit isometries which take the form of integral operators.

For $1 \leq p < \infty$, let H^p denote the Hardy spaces of the disc Δ and let $\|\cdot\|_p$ denote the usual norm. By S^p we will mean the class of analytic functions on Δ for which f' belongs to H^p.

The first norm we consider for S^p is $\|f\| = |f(0)| + \|f'\|_p$. The form of an isometry of H^p plays a significant role in the proof of the following theorem.

4.5.1. THEOREM. *(Novinger and Oberlin) Let T be an isometry of S^p into S^p. Then there exists a linear isometry τ of H^p into H^p and a unimodular complex number λ such that for all $f \in S^p$ and $z \in \Delta$,*

$$Tf(z) = \lambda[f(0) + \int_0^z \tau[f'(\xi)]d\xi].$$

PROOF. Let n be a positive integer and t a real number. We define the function $p_n(z) \equiv z^n$. S^p contains the polynomials and thus $f = 1 + tp_n$ belongs to S^p for every real number t and positive integer n. The fact that, $\|Tf\| = \|f\|$ together with a simple application of the triangle inequality for the norm yields

(58) $$|T1(0) + tTp_n(0)| = |T1(0)| + |t||(Tp_n(0))$$

and

(59) $$\|(T1)' + t(Tp_n)'\|_p = \|(T1)'\|_p + |t|\|(Tp_n)'\|_p.$$

Let n be a positive integer such that Tp_n is a nonconstant function. The real valued function $\rho(t) = \|(T1)' + t(Tp_n)'\|_p$. The last equation above implies that ρ is not differentiable as a function of t at $t = 0$. However the L^p norm is weakly differentiable at every point except the zero vector. Hence, $(T1)'(e^{i\theta}) = 0$ on a set of positive measure if $p = 1$ and almost everywhere if $p > 1$. Since $(T1)' \in H^p$, it follows that $(T1)' \equiv 0$ and therefore must be constant. The constant must be of modulus 1 and without loss of generality we may assume that $T1 = 1$.

It now follows from the first of the equations above that

$$Tp_n(0) = 0$$

for every positive integer n.

For a arbitrary $f \in S^p$, let $g(z) = 1 + t(f(z) - f(0))$. Since $\|Tg\| = \|g\|$ we obtain the following:

$$\begin{aligned}
\|g\| &= 1 + |t| \|f'\|_p \\
\|Tg\| &= \|T1 + tT(f - f(0))\| \\
&= |1 + t(Tf(0) - f(0)| + |t| \|(Tf)'\|_p.
\end{aligned}$$

Thus,

(60) $$1 + |t|(\|f'\|_p - \|(Tf)'\|_p) = |1 + t(Tf(0) - f(0)|.$$

Since the left hand side of this last equation depends on $|t|$, it follows that

(61) $$Tf(0) = f(0) \quad \text{and} \quad \|(Tf)'\|_p = \|f'\|_p.$$

To complete the proof we need to consider a subspace of S^p. Let S_0^p denote the subspace of functions in S^p that vanish at the origin. The differentiation operator D maps S_0^p isometrically onto H^p and its inverse is given by

$$Ig(z) = \int_0^z g(\xi)d\xi.$$

It is clear from the last equation that the isometry T maps the subspace S_0^p onto itself and so we have that the composition $DTI : H^p \to H^p$ is an isometry. Let τ denote this H^p isometry. For an arbitrary $f \in S^p$ it is clear that

$$I(f') = f - f(0).$$

Since $DTI = \tau$ we have

$$D(T(f - f(0))) = \tau(f').$$

This last equation implies that $(Tf)' = \tau(f')$ and so

$$Tf(z) = f(0) + \int_0^z \tau[f'](\xi)d\xi.$$

This completes the proof. □

Now we recall the result of Forelli on isometries to obtain the following corollaries.

4.5.2. COROLLARY. *Let T be an isometry of S^p into S^p and suppose that $p \neq 2$. Then there is a nonconstant inner function ϕ and a function F in H^p such that for $z \in \Delta$ and $f \in S^p$,*

$$Tf(z) = \lambda[f(0) + \int_0^z F(\xi)f'(\phi(\xi))d\xi].$$

4.5.3. COROLLARY. *Let T be an isometry of S^p onto S^p and $p \neq 2$. Then there exist unimodular complex numbers λ, μ and a conformal map ϕ of unit disc Δ such that for $z \in \Delta$ and $f \in S^p$,*

$$Tf(z) = \lambda[f(0) + \mu \int_0^z [\phi'(\xi)]^{1/p} f'(\phi(\xi)) d\xi].$$

Conversely, T defined in this way is an isometry of S^p onto itself.

4.6. Notes and Remarks

The previous sections in this chapter provide the reader with a bare introduction to the isometries of spaces of analytic functions on the disk. The study of Banach spaces of analytic functions goes far beyond the situations we have encountered thus far. At the end of these notes we will attempt to guide the reader to some of the other more interesting results for spaces of functions of several variables.

Some of the most interesting spaces of analytic functions are *function algebras*, closed subalgebras of the Banach space of continuous functions $C(Q)$ on a compact Hausdorff space Q which contain the identity and separate points. These are also called *sup-norm algebras*. The surjective isometries of function algebras were first described by deLeeuw, Rudin, and Wermer [85], and it was their theorem, a version of which we stated and proved as Theorem 2.3.16 in Chapter 2, which led to the isometry theorems on H^1 and H^∞.

Isometries of the Hardy Spaces of the Disk. Theorems 4.2.2 and 4.2.5 were stated and proved in the paper of deLeeuw, Rudin, and Wermer [85] mentioned above. The fact given in Lemma 4.2.1 that every automorphism of H^∞ is induced by a conformal map of the unit disk was first shown by Rudin [272] although our proof mimics the one given in [132, p.144]. As we saw, the proof for H^1 depends upon the the structure of extreme points of $H^1(\Delta)$ given in Theorem 4.2.3 and the form of isometries on $H^\infty(\Delta)$. The proofs of Theorem 4.2.3 and Corollary 4.2.4 were first given by de Leeuw and Rudin [84]. We have chosen not to include the proofs, since they are somewhat lengthy. We refer the reader to the exposition given by Hoffman [132, pp.138-142]. It was noted by de Leeuw, Rudin, and Wermer that after completing their paper [85], they became aware of a paper by Nagasawa [231] which contained Theorems 4.2.2 and 4.2.5 in the context of a study of operator algebras on a Hilbert space.

The results involving H^p for $1 < p < \infty$ where $p \neq 2$ are contained in the paper of Forelli [106] and we have followed his exposition for the most part, although we have also been influenced by the work of Schneider [279] and Fisher [100].

Bergman Spaces. The isometries of the Bergman spaces were first characterized by Kolaski [173] and the proof of Theorem 4.3.1 follows the one given in that paper. We have used an idea of Hornor to avoid the use of Lemma 4.2 in Kolaski's paper. The influence of Rudin [275] is also apparent in Kolaski's work. Kolaski actually stated and proved the theorem for Runge Domains, rather than the open unit disk, and the proof for that more general case requires very little change from the one we have given.

The examples given after the proof of the theorem are due to Kolaski also, and the reader should consult [173] for more details. A detailed study of the nature of the analytic self-maps of the disk which give rise to isometries of the Bergman space has been carried out by Hornor and Jamison [138]. They show, among other things, that for an isometry on $L_a^p(\Delta)$ given by $Tf = h(f \circ \varphi)$, then for each $z \in \Delta$, the set $\varphi^{-1}(\{z\})$ must have constant cardinality a.e. and in case that constant is a finite integer m, the function h is given by $h = \dfrac{\lambda}{m^{1/p}} (\varphi')^{2/p}$ for some unimodular constant λ.

Bloch Spaces. The material in this section is taken exclusively from the paper of Cima and Wogen [72]. They indicate that their work is patterned after a proof in [132, p.141] of a theorem describing isometries of function algebras. Some general properties of the Bloch space can be found in [6] and [71]. We remind the reader that when we say that the isometries on the Bloch space are not canonical, we mean they are not strictly in the form of weighted composition operators.

S^p **Spaces.** The isometries of the S^p spaces were obtained by Novinger and Oberlin in [238]. Their work was motivated by a paper of Roan [267] which was concerned with composition operators of various types on S^p. The form of the isometry obtained in Theorem 4.5.1 is quite interesting. The results in Corollaries 4.5.2 and 4.5.3 follow directly from the characterizations of H^p-isometries by Forelli given in Section 3. Roan had shown that if T is an isometry and a composition operator, then the function on Δ is conformal and takes 0 to 0. Novinger and Oberlin show that Roan's result follows from Corollary 4.5.3. They also consider a second norm on S^p given by

$$\|f\| = \|f\|_\infty + \|f'\|_p.$$

It is then shown in [238] that every isometry on S^p with the above norm is a weighted composition operator.

In a paper which extends some of the above methods, Hornor and Jamison characterize the surjective isometries on the analytic Besov spaces and the Dirichlet spaces [139].

Further remarks: several variables and related matters. Theorem 4.2.2 has an immediate application to the space $H^\infty(B^n)$. We follow Rudin [276]. Recall that the ball in the n dimensional Hilbert space C^n is the set

$$B = \left\{ (z_1, z_2, .., z_n) : \left(\sum_{j=1}^{j=n} |z_j|^2 \right)^{1/2} < 1 \right\}.$$

Let $Aut(B)$ denote the automorphisms of B. For a fixed $a \in B$, let P_a be the orthogonal projection of C^n onto the subspace $[a]$ generated by a, and let

$Q_a = I - P_a$. Thus $P_0 = 0$ and for $a \neq 0$

$$P_a = \frac{<z, a>}{<a, a>} a.$$

Set $s_a = (1 - |a|^2)^{1/2}$ and define

$$\phi_a(z) = \frac{a - P_a z - s_a Q_a z}{1 - <z, a>}.$$

From Rudin [276] we quote

4.6.1. THEOREM. *If* $\psi \in Aut(B)$ *and* $a = \psi^{-1}(0)$ *then there exists a unique* $n \times n$ *unitary matrix* U *such that*

$$\psi = U\phi_a.$$

The identity

$$1 - <\psi(z), \psi(w)> = \frac{(1 - <a, a>)(1 - <z, w>)}{(1 - <z, a>)(1 - <a, w>)}$$

holds for all $z \in \overline{B}$, $w \in \overline{B}$.

Also from Rudin [276] we have

4.6.2. THEOREM. *If* T *is an automorphism of the algebra* H^∞, *then* $Tf = f \circ \psi$ *for some* $\psi \in Aut(B)$.

The following corollary is now clear.

4.6.3. COROLLARY. *If* T *is a linear isometry of* $H^\infty(B)$ *onto* $H^\infty(B)$, *or an isometry of* $A(B)$ *onto* $A(B)$, *then there is a* $\alpha \in C$ *with* $|\alpha| = 1$ *and a* $\psi \in Aut(B)$ *such that*

$$Tf = \alpha f \circ \psi$$

The results of Forelli for $H^p(\Delta)$ provided the catalyst for several investigators to attempt to extend his results to other settings. Schneider[279] and Forelli [107] were the first to consider spaces of several variables and both gave a version of Forelli's $H^p(\Delta)$ result for the space $H^p(U^n)$. Both of these results are for the surjective case and were restricted to $p > 2$. Rudin [275] and [276] settled the surjective case completely with the following theorem.

4.6.4. THEOREM. *Suppose* $0 < p < \infty$, $p \neq 2$, $n \geq 1$, *and* T *is a linear isometry of* $H^p(B)$ *onto* $H^p(B)$. *Then there is a* $\psi \in Aut(B)$ *and a* $c \in C$, $|c| = 1$, *such that*

$$(Tf)(z) = c \frac{(1 - |a|^2)^{n/p}}{(1 - <z, a>)^{2n/p}} f(\psi(z))$$

for all $f \in H^p(B)$, $z \in B$, *where* $a = \psi^{-1}(0)$.

The methods of Rudin were extended to the case of H^p spaces of bounded symmetric domains by Koranyi and Vagi [183]. Kolaski [173],[174], and [176] showed that the methods of Forelli and Rudin also applied to the weighted Bergmann spaces of bounded symmetric domains. Recently the Rudin method was applied by Garling and Wojtaszczyk to obtain a characterization of the surjective isometries of the Bargmann spaces [111].

The work of Cima and Wogen has also been extended to the several variable setting. In particular Krantz and Ma [[184] prove the following:

4.6.5. THEOREM. *Let* $U : \mathcal{B}_0(B^n) \to \mathcal{B}_0(B^n)$ *be an isometric isometry. Then there is a* $\phi \in Aut(B^n)$ *and a* $\mu \in \mathbf{C}$ *with* $|\mu| = 1$ *with*

$$Uf = \mu(f(\phi(z)) - f(\phi(0)))$$

for every $f \in \mathcal{B}_0(B^n)$.

Kolaski [175] extended the methods of Novinger and Oberlin [238] to obtain the following nice generalization of their results. First some notation. Let Δ denote the open unit disk and $H(\Delta)$ be the analytic functions on Δ. Let $N : H(\Delta) \to [0, \infty]$ be a norm on $H(\Delta)$ Let H_N and S_N denote the spaces of functions in $H(\Delta)$ such that $N(f) < \infty$ and $N(f') < \infty$, respectively. Let H_N be given the norm N, and give S_N the norm

$$\|f\| = |f(0)| + N(f').$$

4.6.6. THEOREM. *(Kolaski) Let* T *be an isometry of* S_N *into(onto)* S_N. *If* H_N *is smooth, or if* $T1$ *is a constant function, then there is a linear isometry* A *of* H_N *into(onto)* H_N, *and a* $\lambda \in \Gamma$ *such that*

$$Tf(z) = \lambda \left(f(0) + \int_0^z A(f')(\xi)d\xi \right)$$

for $f \in S_N$, $z \in \Delta$.

Other papers which are related to what has been discussed in this chapter include [3], [26], [27], [28], [29], [30], [31], [32], [33], [53], [93], [94], [95], [149], [192], [215], [216], [217], [218], [221], [226], [293], [291], and [301].

CHAPTER 5

Rearrangement Invariant Spaces

5.1. Introduction

In the previous chapters we have focused our attention on some specific classical Banach spaces. In this chapter, we consider a more general class of spaces which includes the L^p-spaces as a special case. Already we have seen a step in that direction in a theorem of Lamperti (mentioned as Theorem 3.5.1 in Chapter 3) which is a partial result for Orlicz spaces. At the suggestion of Lamperti, Lumer extended the result to reflexive Orlicz spaces using a method now referred to as "Lumer's method" and which has been used extensively by many authors in a variety of settings. This method, first mentioned in the Remarks at the end of Chapter 1, will be prominent in the proofs given in this chapter, and we will develop it fully.

For the most part, the classes of spaces under consideration in the present chapter can be included under the general heading of *Banach function spaces*. These spaces consist of measurable functions on some measure space and on which there is a norm related to the underlying measure in a certain way so that the resulting space is a Banach space. To be more precise, let us give some definitions.

5.1.1. DEFINITION. *Let (Ω, Σ, μ) be a measure space. A mapping ρ from the nonnegative σ-measurable functions into $[0, \infty]$ is called a* function norm *if for all nonnegative measurable functions f, g, f_n, $(n = 1, 2, 3, \ldots)$, for all constants $a \geq 0$, and for all $E \in \Sigma$, the following properties hold:*

(i) $\rho(f) = 0 \Leftrightarrow f = 0 \quad \mu\text{-}a.e.$,
(ii) $\rho(af) = a\rho(f)$,
(iii) $\rho(f + g) \leq \rho(f) + \rho(g)$,
(iv) $0 \leq g \leq f \quad \mu\text{-}a.e. \Rightarrow \rho(g) \leq \rho(f)$,
(v) $0 \leq f_n \nearrow f \quad \mu\text{-}a.e. \Rightarrow \rho(f_n) \nearrow \rho(f)$,
(vi) $\mu(E) < \infty \Rightarrow \rho(\chi_E) < \infty$,
(vii) $\mu(E) < \infty \Rightarrow \int_E f d\mu \leq C_E \rho(f)$

for some constant C_E, $0 < C_E < \infty$, depending on E and ρ but independent of f.

5.1.2. DEFINITION. *Let ρ be a function norm. The collection $X = X(\rho)$ of all Σ-measurable functions f for which $\rho(|f|) < \infty$ is called a* Banach function space. *For each $f \in X$, define*

$$\|f\|_X = \rho(|f|).$$

A function f is said to have absolutely continuous norm *in X if $\|f\chi_{E_n}\|_X \to$
0 for every sequence $\{E_n\}$ of measurable sets for which the characteristic
functions χ_{E_n} converge to 0 μ-a.e. If this is true for every $f \in X$, then X is
said to have* absolutely continuous norm.

The most obvious examples of Banach function spaces are, of course, the
L^p spaces and these spaces and their properties provide the model for the
general case. The "modular" (34) of Lamperti as defined in Chapter 3 and
the corresponding space L^φ discussed there gives a generalization of the L^p
spaces, and Theorem 3.5.1 mentioned above characterizes those isometries
which preserve the modular. This result was the motivation for Lumer to
consider the problem of determining the form of all isometries on general
Orlicz spaces. The canonical form given by Lamperti does define an operator
which preserves the modular and so is an isometry. The converse was not so
clear; that is, must every isometry be a modular isometry?

In Section 2, we will give Lumer's work in which he shows that any surjective isometry on a complex reflexive Orlicz space over a nonatomic measure
space and where the space is equipped with the so-called *Luxemburg norm*, is
in the canonical form. Here he describes the form of the semi-inner product
on the space which gives the norm, and characterizes the Hermitian operators.
From the form of the Hermitians, he is then able to determine the isometries.

Zaidenberg used Lumer's approach to remove the conditions of reflexivity
and separability, and in fact extended the result to the larger class of *symmetric* or *rearrangement invariant* spaces. We will give an account of this work
in Section 3.

In the final section we generalize one step further to determine the surjective isometries on certain *Musielak-Orlicz* spaces. There are two norms
commonly defined on a given Musielak-Orlicz space (as is also the case for
Orlicz spaces) called the *Luxemburg norm* and the *Orlicz norm* and these
norms are equivalent. We will end by giving Kaminska's proof that the groups
of isometries for the two norms coincide.

5.2. Lumer's Method for Orlicz Spaces

It is reasonable to begin with some definitions and terminology in regard
to Orlicz spaces, even though these will be included in the more general definitions given in the section on Musielak-Orlicz spaces. In what follows, we
will assume the measure space (Ω, Σ, μ) to be σ-finite.

5.2.1. DEFINITION. *Let φ be a nonnegative, increasing, convex function
with $\varphi(0) = 0$. Let $I_\varphi(f) \equiv \int_\Omega \varphi(|f|)d\mu$. The set L^φ of all Σ-measurable
functions f for which $I_\varphi(rf) < \infty$ for some $r > 0$ is called an* Orlicz space.
A function φ with the above properties is called a Young's function.

The space L^φ as defined above is slightly different than the one given in Chapter 3, which is not necessarily a linear space. The two are the same in case φ satisfies the Δ_2-condition.

5.2.2. DEFINITION.

(i) *A Young's function φ is said to satisfy the Δ_2-condition if there exist $s_0 > 0$ and $c > 0$ such that*

$$\varphi(2s) \le c\varphi(s) < \infty, \quad (s_0 \le s < \infty).$$

(ii) *A Young's function φ is said to satisfy the δ_2-condition if there exist $s_0 > 0$ and $c > 0$ such that $\varphi(t) > 0$ for all $t > 0$, and*

$$\varphi(2t) \le c\varphi(t) \ for \ 0 \le t \le s_0.$$

If φ is a Young's function, then the function φ^* defined by

(62)
$$\varphi^*(u) = \sup_{v \ge 0}\{uv - \varphi(v)\}$$

is also a Young's function called the *conjugate* or *associate* function to φ, and the two functions φ, φ^* are said to be *complementary* Young's functions.

For f in L^φ we define

(63)
$$\|f\|_l = \inf\{\epsilon > 0 : I_\varphi(\frac{f}{\epsilon}) \le 1\}$$

and

(64)
$$\|f\|_O = \sup\{\int |fg| : I_{\varphi^*}(g) \le 1\}.$$

Each of the above equations defines a Banach function norm for which the corresponding Banach function space is the Orlicz space L^φ. The norm $\|\cdot\|_l$ in (63) is called the *Luxemburg* norm on L^φ while $\|\cdot\|_O$ given by (64) is called the *Orlicz* norm. We will drop the subscripts when the context makes clear which norm we mean.

5.2.3. EXAMPLE.

(i) *If $1 \le p < \infty$, and $\varphi(t) = t^p$, then φ is a Young's function and the corresponding Orlicz space is L^p.*

(ii) *If φ is defined by*

$$\varphi(t) = \begin{cases} 0, & if \ 0 \le t \le 1; \\ \infty & if \ t > 1, \end{cases}$$

then $L^\varphi = L^\infty$ and the set of f such that $I_\varphi(f) \le 1$ is the unit ball of L^∞.

(iii) *If φ is defined by*

$$\varphi(t) = \begin{cases} 0, & \text{if } 0 \leq t \leq 1; \\ t\log t & \text{if } t > 1, \end{cases}$$

then L^φ is the Zygmund space $L\log L$.

We note that the Young's functions in (i) and (iii) above satisfy the Δ_2-condition, while the function in (ii) does not. In this latter case, the Orlicz class as defined by Lamperti in Chapter 3, is not a linear space although it is convex.

It is convenient now to gather together some facts about Orlicz norms and Orlicz spaces for future reference. We omit the proofs.

5.2.4. THEOREM. *(Properties of Orlicz spaces)*

(i) *If φ is a Young's function, then the Luxemburg and Orlicz norms on L^φ are equivalent:*

$$\|f\|_l \leq \|f\|_O \leq 2\|f\|_l.$$

(ii) *If φ is a Young's function, then L^φ is the closure of the integrable simple functions if the norm on L^φ is absolutely continuous.*

(iii) *If φ is a Young's function and φ^* is the conjugate function, then $(L^\varphi)^* = L^{\varphi^*}$ if and only if the norm on L^φ is absolutely continuous.*

(iv) *If φ is a Young's function and φ^* is the conjugate function, then L^φ is reflexive if and only if the norms on both L^φ and L^{φ^*} are absolutely continuous.*

(v) *If φ is a Young's function, then L^φ is separable if and only if the norm is absolutely continuous and the measure μ is separable (that is, if the σ-algebra Σ has a countable set of generators).*

(vi) *If φ is a Young's function then the norm on L^φ is absolutely continuous if and only if*
    (a) *$\mu(\Omega) < \infty$ and φ satisfies the Δ_2-condition, or*
    (b) *$\mu(\Omega) = \infty$, there is set of positive measure which is free of atoms, and φ satisfies both the Δ_2 and δ_2-conditions, or*
    (c) *$\mu(\Omega) = \infty$, the measure space is purely atomic with bounded masses, and φ satisfies the δ_2-condition.*

(vii) *If φ is a Young's function and the norm on L^φ is absolutely continuous, then $I_\varphi(rf) < \infty$ for all $r > 0$.*

(viii) *If φ is a Young's function and φ^* is the conjugate function then*
    (a) *$st \leq \varphi(s) + \varphi^*(t)$, (Young's inequality),*
    (b) *$s\varphi'(s) = \varphi(s) + \varphi^*(\varphi'(s))$, (Young's equality),*
    (c) *$t\varphi'(t) \leq \varphi(2t)$ for all $t > 0$.*

The function φ' above will denote the left derivative of φ, which is left continuous and is the actual derivative of φ almost everywhere. The condition in (viii)a above is known as Young's inequality and (viii)b shows that equality occurs when $t = \varphi'(s)$.

We remarked in the notes to Chapter 1 that one method of finding isometries on a space was to first find the Hermitian operators. This is what we call *Lumer's method*, and we will follow that approach in this section. If $[\cdot, \cdot]$ is a semi-inner product on a Banach space X which is compatible with the norm on X, (see 1.4.1 in Chapter 1) then a bounded operator H on X is *Hermitian* if and only if $[Hx, x]$ is real for all $x \in X$. This definition only makes sense if the space is complex, and we will assume that for the remainder of this section. It is also important to note that, although there may not be a unique semi-inner product that is compatible with the norm, the definition as given does not depend on that particular semi-inner product. For, if we let

$$(65) \qquad W[T] = \{[Tx, x] : \|x\| = 1\}$$

denote the *numerical range* of the operator T with respect to the given s.i.p., then by a result of Lumer, all determinations of the numerical range of an operator have the same convex hull.

There are some other equivalent properties that characterize Hermitian operators and it will be useful to have a clear statement of those available to us. For the proof of these equivalencies, we will need the following lemma.

5.2.5. LEMMA. *For any bounded operator T on a Banach space X and any s.i.p. $[\cdot, \cdot]$ compatible with the norm on X, let*

$$\beta = sup\{\Re\lambda : \lambda \in W(T)\}.$$

Then

(i) $\beta = \lim\limits_{t \to 0^+} \dfrac{\|I + tT\| - 1}{t}$;

(ii) $\beta = \sup\{\dfrac{1}{t} \log \| \exp(tT)\| : t > 0\}$;

(iii) $\beta = \lim\limits_{t \to 0^+} \dfrac{1}{t} \log \| \exp(tT)\|$.

PROOF. Let $\delta(T)$ denote the limit in (i), which exists since $\|I + tT\|$ is a convex function of t. If $x \in X$ with $\|x\| = 1$ and $t > 0$, then

$$\begin{aligned}
\|(I + tT)x\| &\geq |[(I + tT)x, x]| \\
&= |1 + t[Tx, x]| \\
&= \sqrt{(1 + t\Re[Tx, x])^2 + (t\Im[Tx, x])^2} \\
&\geq \sqrt{1 + 2t \inf \Re(W(T))}
\end{aligned}$$

for t sufficiently small. For such t and any $x \in X$ we have

$$(66) \qquad \|(I + tT)x\| \geq \sqrt{1 + 2t \inf \Re(W(T))}\|x\|.$$

For $|t| < 1/\|T\|$, $F(t) = (I + tT)^{-1}$ exists and if t is also sufficiently small that (66) holds, then

$$\|F(t)\| \leq \frac{1}{\sqrt{1 + 2t \inf \Re(W(T))}}.$$

Since

$$F(t) = \sum_{n=0}^{\infty} (-tT)^n = I - tT + t^2 T^2 (F(T)),$$

we have

$$|\|I - tT\| - \|F(t)\|| \leq t^2 \|T\|^2 \|F(t)\|.$$

The right hand side goes to zero as $t \to 0^+$ and we get

$$\lim_{t \to 0^+} \|I - tT\| = \lim_{t \to 0^+} \|F(t)\|.$$

Using these facts we obtain

$$\delta(-T) = \lim_{t \to 0^+} \frac{\|F(t)\| - 1}{t}$$

$$\leq \lim_{t \to 0^+} \frac{1}{t} \left(\frac{1}{\sqrt{1 + 2t \inf \Re(W(T))}} - 1 \right)$$

$$= - \inf \Re(W(T)).$$

Now replace T by $-T$ to get

$$\delta(T) \leq - \inf \Re(T) = \sup \Re(W(T)) = \beta.$$

For $\|x\| = 1$, we have

$$\Re[Tx, x] = \lim_{t \to 0^+} \frac{1}{t} (\sqrt{1 + 2t \Re[Tx, x] + t^2 |[tx, x]|^2} - 1)$$

$$\leq \lim_{t \to 0^+} \frac{1}{t} (\|I + tT\| - 1).$$

Thus $\beta \leq \delta(T)$ and (i) is proved.

For any $x \in X$ and sufficiently small $t > 0$, we have from (66) that

$$\|(I - tT)x\| \geq (\sqrt{1 + 2t \inf \Re(W(-T))}) \|x\|$$

$$= (\sqrt{1 - 2t\beta}) \|x\|$$

$$\geq (1 - t\beta) \|x\|.$$

By induction, we get

$$\|(I - tT)^n x\| \geq (1 - t\beta)^n \|x\|$$

so that for any $t > 0$ and sufficiently large n we have

$$\|(I - \frac{t}{n} T)^n x\| \geq (1 - \frac{t}{n} \beta)^n \|x\|.$$

Letting n go to ∞ in the inequality above, we obtain

$$\|exp(-tT)x\| \geq exp(-t\beta) \|x\|.$$

It follows by letting $x = \exp(tT)y$, where $\|y\| = 1$, that

$$1 = \|y\| \geq \exp(-t\beta)\|\exp(tT)y\|$$

from which we conclude that $\|\exp(tT)\| \leq \exp(t\beta)$ for all $t > 0$. Hence,

(67) $$\sup_{t>0}\frac{1}{t}\log\|\exp(tT)\| \leq \beta.$$

We may write $\|\exp(tT)\| = \|I + tT\| + f(t)$ where $|f(t)| \leq \gamma t^2$ for some $\gamma > 0$ and $0 \leq t \leq 1$. Since $\log t \geq \dfrac{t-1}{t}$ for $t > 0$, we see that

$$\frac{1}{t}\log\|\exp(tT)\| \geq \frac{\frac{1}{t}(\|I + tT\| - 1) + \frac{1}{t}f(t)}{\|I + tT\| + f(t)}.$$

Now, we let $t \to 0^+$ to obtain

(68) $$\lim_{t\to 0^+}\frac{1}{t}\log\|\exp(tT)\| \geq \beta.$$

Both (ii) and (iii) follow now from (67) and (68).   □

5.2.6. THEOREM. *Let H be a bounded operator on a Banach space X. The following are equivalent:*

(i) *H is Hermitian; that is, $W(H) \subset \mathbb{R}$.*
(ii) *$\|I + itH\| = 1 + o(t)$.*
(iii) *$\|\exp(itH)\| = 1$ for all $t \in \mathbb{R}$.*
(iv) *$\exp(itH)$ is an isometry for each $t \in \mathbb{R}$.*

PROOF. We note that $W(H) \subset \mathbb{R}$ implies that

(69) $$\sup \Re(W(iH)) = \sup \Re(W(-iH)) = 0$$

and by part (i) of Lemma 5.2.5 we must have

$$\lim_{t\to 0}\frac{1}{t}(\|I + itH\| - 1) = 0.$$

The equivalence of (i) and (ii) in the theorem follows from this.

Again suppose that H is Hermitian. Then (69) holds and by Lemma 5.2.5 (ii) applied to both H and $-H$, we must have $\|\exp(itH)\| \leq 1$ for all real t. It follows from this that $\|\exp(itH)\| = 1$ for all real t. Thus (i) implies (iii) in the statement of the theorem.

On the other hand, if $\|\exp(itH)\| = 1$ for all $t \in \mathbb{R}$, then (69) is satisfied by Lemma 5.2.5, and so we have $\|I + itH\| = 1 + o(t)$ as before.

Finally, if part (iii) of the theorem holds, then for any $x \in X$,

$$\|x\| = \|\exp(-itH)(\exp(itH)x)\| \leq \|\exp(itH)x\| \leq \|x\|$$

so that (iii) implies (iv). Since the other implication is obvious, the proof is complete.   □

From here on we shall assume that the Young's function φ is such that the norm on L^φ is absolutely continuous. (See Theorem 5.2.4(vi).) In particular, this means that if $f \in L^\varphi$, then $\int \varphi(r|f|)d\mu < \infty$ for every $r > 0$.

The first thing we will do is describe a semi-inner product which is compatible with the Luxemburg norm on an Orlicz space.

5.2.7. LEMMA. *If the norm on L^φ is absolutely continuous, then $\|f\| = 1$ if and only if $I_\varphi(f) = 1$.*

PROOF. Suppose $\|f\| = 1$. Then $\int \varphi\left(\frac{|f|}{\epsilon}\right) d\mu \leq 1$ for every $\epsilon > 1$. If $\{\epsilon_n\}$ is a sequence approaching 1 from above, then

$$\int \varphi\left(\frac{|f|}{\epsilon_n}\right) d\mu \to \int \varphi(|f|)d\mu$$

by the Monotone Convergence Theorem, and we must have $I_\varphi(f) \leq 1$.

On the other hand, if $\{\epsilon_n\} \to 1$ from below, then $\int \varphi\left(\frac{|f|}{\epsilon_n}\right) d\mu > 1$ for each n, $I_\varphi\left(\frac{f}{\epsilon_1}\right) < \infty$, and by the Lebesgue Convergence Theorem, we must have $I_\varphi(f) \geq 1$.

For the converse, note that if $I_\varphi(f) = 1$, then $\|f\| \leq 1$. Therefore we must have

$$1 = \int \varphi(|f|)d\mu \leq \int \varphi\left(\frac{|f|}{\|f\|}\right) d\mu = 1$$

and hence,

$$\varphi(|f|/\|f\|) = \varphi(|f|) \quad a.e.$$

If $(|f(t)|/\|f\|) = (|f(t)|)$ for even one $t \in \Omega$ where $|f(t)| \neq 0$, then $\|f\| = 1$. If this fails, then $|f|/\|f\|$ has all its values in an interval of constancy for φ. This would lead, by Young's equality 5.2.4(viii)b to

$$(70) \quad 0 = \frac{|f|}{\|f\|}\varphi'\left(\frac{|f|}{\|f\|}\right) = \varphi\left(\frac{|f|}{\|f\|}\right) + \varphi^*\left(\varphi'\left(\frac{|f|}{\|f\|}\right)\right) = \varphi\left(\frac{|f|}{\|f\|}\right) \quad a.e.$$

which is impossible.   □

5.2.8. LEMMA. *Let $L^\varphi(\Omega, \Sigma, \mu)$ be an Orlicz space with absolutely continuous norm. For $f, g \in L^\varphi$, let*

$$(71) \qquad\qquad [f, g] = C(g) \int f\varphi'\left(\frac{|g|}{\|g\|}\right) sgn g d\mu$$

where

$$C(g) = \|g\|^2 / \int |g|\varphi'\left(\frac{|g|}{\|g\|}\right) d\mu.$$

Then $[f, g]$ defines a semi-inner product on L^φ which is compatible with the norm.

PROOF. We interpret $C(g)$ to be 0 when $g = 0$ a.e. The integral in the denominator of the expression for $C(g)$ is positive unless $g = 0$ a.e. That this is so can be seen from replacing f by g in (70) above. To see that the integral is finite, we use the inequality (viii)c from Theorem 5.2.4. Thus

$$\int |g|\varphi'\left(\frac{|g|}{\|g\|}\right) d\mu = \|g\| \int \frac{|g|}{\|g\|}\varphi'\left(\frac{|g|}{\|g\|}\right) d\mu \leq \|g\| \int \varphi\left(\frac{2|g|}{\|g\|}\right) d\mu < \infty.$$

It is clear from the definitions that $[g,g] = \|g\|^2$ for any $g \in L^\varphi$. Hence $[g,g]$ is positive definite, and $[f,g]$ is obviously linear in the left argument. To complete the proof, we must show that $|[f,g]| \leq \|f\|\|g\|$ for all f,g in L^φ.

Suppose that f,g are each of norm 1 so that $\int \varphi(|f|)d\mu = \int \varphi(|g|)d\mu = 1$ by Lemma 5.2.7. Then

$$|[f,g]| = C(g)\left|\int f\varphi'\left(\frac{|g|}{\|g\|}\right) sgn g\, d\mu\right|$$

$$\leq C(g)\int |f|\left|\varphi'\left(\frac{|g|}{\|g\|}\right)\right| d\mu$$

$$\leq C(g)\int \left\{\varphi(|f|) + \varphi^*\left(\varphi'\left(\frac{|g|}{\|g\|}\right)\right)\right\} d\mu \quad \text{(by Theorem 5.2.4(viii)a)}$$

$$= \frac{\int\{\varphi(|f|) + \varphi^*(\varphi'(|g|))\}d\mu}{\int\{\varphi(|g|) + \varphi^*(\varphi'(|g|))\}d\mu} \quad \text{(by Theorem 5.2.4(viii)b)}$$

$$= 1.$$

The desired result now comes easily since positive scalars factor out of both components in $[f,g]$.   □

Next, we establish a class of operators which are Hermitian on L^φ.

5.2.9. LEMMA. *If $h \in L^\infty(\Omega, \Sigma, \mu)$ is real valued, the operator H_h defined by $H_h f = hf$ for all $f \in L^\varphi$ is a bounded Hermitian operator, and $\|H_h\| = \|h\|_\infty$.*

PROOF. For the s.i.p. $[\cdot, \cdot]$ as given by (71), we have

$$[H_h f, f] = C(f)\int_\Omega hf\varphi'\left(\frac{|f|}{\|f\|}\right) sgn f\, d\mu = C(f)\int_\Omega h|f|\varphi'\left(\frac{|f|}{\|f\|}\right) d\mu$$

is real for all f.

Since $|hf| \leq \|h\|_\infty|f|$ a.e., we have

$$\|H_h f\| \leq \|h\|_\infty\|f\|$$

for all f, and so $\|H_h\| \leq \|h\|_\infty$. Here we have used the fact that for any $f,g \in L^\varphi$, $|f| \leq |g|$ implies that $\|f\| \leq \|g\|$. On the other hand, given $\epsilon > 0$, there exists $E \in \Sigma$ with $\mu(E) > 0$ such that $|h(t)| \geq \|h\|_\infty - \epsilon$ for all $t \in E$. For $f = \chi_E/\|\chi_E\|$ we obtain

$$\|H_h f\| = \|hf\| \geq (\|h\|_\infty - \epsilon)$$

from which we conclude that $\|H_h\| \geq \|h\|_\infty$.   □

It is now the goal to show that every Hermitian operator is of the form given in the above lemma.

5.2.10. LEMMA. *(Lumer) Let H be any bounded Hermitian operator on L^φ and suppose that f, g are nonzero functions whose supports E_1, E_2 are such that $\mu(E_1 \cap E_2) = 0$. Then*

$$(72) \qquad \int_{E_2} Hf\varphi'\left(\frac{|g|}{\|f+g\|}\right) sgn g d\mu = \overline{\int_{E_1} Hg\varphi'\left(\frac{|f|}{\|f+g\|}\right) sgn f d\mu}.$$

PROOF. Since f and g have disjoint supports, $\|f + g\| = \|e^{i\alpha}f + e^{i\beta}g\|$ for all real numbers α, β. Since H is Hermitian, $[H(e^{i\alpha}f + e^{i\beta}g), e^{i\alpha}f + e^{i\beta}g]$ is real and some manipulation with the form of the s.i.p. given by (71) leads to the fact that

$$\int_{E_1} Hf\varphi'\left(\frac{|f|}{\|f+g\|}\right) sgn f d\mu + \int_{E_2} Hg\varphi'\left(\frac{|g|}{\|f+g\|}\right) sgn g d\mu$$

$$+ e^{i(\alpha-\beta)} \int_{E_2} Hf\varphi'\left(\frac{|g|}{\|f+g\|}\right) sgn g d\mu$$

$$+ e^{-i(\alpha-\beta)} \int_{E_1} Hg\varphi'\left(\frac{|f|}{\|f+g\|}\right) sgn f d\mu$$

is real for all α, β. If an expression $\gamma + e^{i\theta}\delta + e^{-i\theta}\eta$ is real for all choices of θ, then γ is real and $\delta = \bar{\eta}$. Equation (72) follows from this observation. $\square$

5.2.11. THEOREM. *(Lumer) Suppose H is a bounded Hermitian operator on $X = L^\varphi(\Omega, \Sigma, \mu)$ where the norm is absolutely continuous, and the measure space is purely nonatomic. Then either $X = L^2(\mu)$, or else there exists $h \in L^\infty(\mu)$ such that $Hf = hf$ for all $f \in X$. In this case, $\|H\| = \|h\|_\infty$.*

PROOF. We assume that $\mu(\Omega) < \infty$ and that $E_1 \in \Sigma$. Let χ_1 denote χ_{E_1} and suppose that $H\chi_1$ is not identically zero (a.e.) on $\Omega \backslash E_1$. Then there exists a measurable subset E_2 of $\Omega \backslash E_1$ with $\mu(E_2) > 0$ such that $\int_{E_2} H\chi_1 d\mu \neq 0$. Let $\chi_2 = \chi_{E_2}$ and let α be a non-negative real number. Application of Lemma 5.2.10 yields

$$(73) \qquad \int_{E_1} H(\alpha\chi_2)\varphi'\left(\frac{|\chi_1|}{\|\alpha\chi_2+\chi_1\|}\right) d\mu = \overline{\int_{E_2} H\chi_1\varphi'\left(\frac{|\alpha\chi_2|}{\|\alpha\chi_2+\chi_1\|}\right) d\mu}.$$

This holds for any α; in particular, it holds for $\alpha = 1$ from which we can conclude that

$$(74) \qquad \int_{E_1} H\chi_2 d\mu = \overline{\int_{E_2} H\chi_1 d\mu}.$$

Combining the above equality with (73), we obtain

$$\left\{\alpha\varphi'\left(\frac{1}{\|\alpha\chi_2+\chi_1\|}\right) - \varphi'\left(\frac{\alpha}{\|\alpha\chi_2+\chi_1\|}\right)\right\} \int_{E_1} H\chi_2 d\mu = 0$$

and since the integral is not zero, we must have

$$\alpha\varphi'\left(\frac{1}{\|\alpha\chi_2 + \chi_1\|}\right) = \varphi'\left(\frac{\alpha}{\|\alpha\chi_2 + \chi_1\|}\right).$$

Since the measure is nonatomic, we may let $\|\chi_2\| \to 0$ and therefore obtain the equation

$$\varphi'\left(\frac{\alpha}{\|\chi_1\|}\right) = \alpha\varphi'\left(\frac{1}{\|\chi_1\|}\right)$$

for all $\alpha > 0$. This latter statement would imply that φ' is homogeneous, and so $\varphi(t) = \beta t^2$ for some constant β. In this case, $L^\varphi = L^2(\Omega, \Sigma, \mu)$. Therefore, if we assume that $L^\varphi \neq L^2$, the above calculations lead to a contradiction, from which we are forced to conclude that support$(H\chi_1) \subseteq$ support(χ_1).

Let 1 denote the function that is identically one. Since $\chi_1 - 1 = 0$ on E_1, it is true that $H(\chi_1 - 1) = 0$ on E_1. It follows that $H\chi_1 = H(1)\chi_1$. We let $h = H(1)$, and the extension of the previous equation by linearity to the simple functions is immediate.

A straightforward calculation shows that if E is in Σ with $\mu(E) > 0$, then

$$[H\chi_E, \chi_E] = \frac{\|\chi_E\|^2}{\mu(E)}\int_E h\,d\mu.$$

Since the left side is always real, we see that h must be real valued. Another consequence of the equality above is that $\|H\|$ is a bound for $\frac{1}{\mu(E)}\left|\int_E h\,d\mu\right|$, from which it follows that $h \in L^\infty$.

If $f \in L^\varphi$ and $\{\sigma_n\}$ is a sequence of simple functions converging in norm to f, then $h\sigma_n \to hf$ since h is in L^∞. The density of the simple functions and the continuity of H together show that $Hf = hf$.

In the case where (Ω, Σ, μ) is σ-finite, $\Omega = \bigcup\Omega_n$ where $\mu(\Omega_n) < \infty$, and a bounded Hermitian H can be restricted to H_n on $L^\varphi(\Omega_n)$. By the previous part, $H_n f = h_n f$ where $\|h_n\|_\infty = \|H_n\| \leq \|H\|$. Since h_{n+1} is an extension of h_n, we can define $h(t) = h_n(t)$ if $t \in \Omega_n$. Then h is a real L^∞ function and the fact that $Hf = hf$ follows in the standard way. $\square$

We come now to the main theorem in this section.

5.2.12. THEOREM. *(Lumer) Let* $X = L^\varphi(\Omega, \Sigma, \mu)$ *and let* U *denote an isometry from* X *onto itself, where* X *has an absolutely continuous norm and the measure space is purely nonatomic. Then there exist a regular set isomorphism* T *and a fixed function* $h \in X$ *such that*

(75) $$Uf(t) = h(t)T_1(f)(t)$$

where T_1 *is the operator induced by* T.

PROOF. Again we first assume that the measure space is finite. Let U be a surjective isometry as in the hypotheses. It is not difficult to show that if $[f, g]$ denotes the s.i.p. as given by (71), then

$$[f, g]_U = [U^{-1}f, U^{-1}g]$$

also defines a s.i.p. which is compatible with the norm of X. If H is a Hermitian operator on X, then

$$[UHU^{-1}f, f]_U = [HU^{-1}f, U^{-1}f] \in \mathbb{R}$$

for all $f \in X$, and we have that UHU^{-1} is also Hermitian. Given a real valued function $g \in L^\infty$, it is a consequence of Theorem 5.2.11 that there must be a real L^∞-function $\hat{g}$ such that

$$UgU^{-1}f = UH_gU^{-1}f = \hat{g}f$$

for all $f \in X$. Furthermore, we observe that

$$\widehat{g_1g_2}f = U(g_1g_2)U^{-1}f = Ug_1U^{-1}Ug_2U^{-1}f = \hat{g}_1\hat{g}_2f$$

so that $\widehat{g_1g_2} = \hat{g}_1\hat{g}_2$. Upon applying this fact to the characteristic function of a measurable set A, we conclude that $\widehat{\chi_A}$ has only the values 0 and 1, and must therefore be a characteristic function itself. This establishes a mapping T on Σ defined by

$$(76) \qquad\qquad \widehat{\chi_A} = \chi_{TA}.$$

The set map T defined by (76) is readily shown to be a regular set isomorphism. It is clear that $T(\Omega\backslash A) = T\Omega\backslash TA$. If $\{A_n\}$ is a disjoint sequence of measurable sets, then by absolute continuity of the norm,

$$\widehat{\chi_{\cup A_n}} = \lim_{n\to\infty} \widehat{\chi_{\cup_1^n A_j}}$$

from which we get $T(\cup A_n) = \cup T(A_n)$. Furthermore, if $\mu(TA) = 0$, then $\chi_{TA} = \widehat{\chi_A}$ a.e., and since $\|\widehat{\chi_A}\|_\infty = \|\chi_A\|_\infty = \|H_{\chi_A}\|$, we must have $\mu(A) = 0$.

Let T_1 be defined on the characteristic functions by $T_1(\chi_A) = \chi_{TA}$ and let $h = U(1)$. Then for any $A \in \Sigma$,

$$U\chi_A = U\chi_A 1 = U\chi_A U^{-1}U(1) = \chi_{TA}U1 = hT_1\chi_A.$$

Hence, (75) is satisfied for characteristic functions. As in the proof of Lamperti's Theorem in Chapter 3, we can easily extend T_1 by linearity to the simple functions so (75) holds for those as well. We extend the definition of T_1 by (36) of Chapter 3, so that it possesses the properties outlined there in Remark 3.2.4 (v). In particular, T_1 is a positive map that preserves almost everywhere convergence. Since the set isomorphism T preserves disjointness, it is clear that $|T_1(\sigma)| = T_1(|\sigma|)$ for each simple function σ and therefore, $|T_1(f)| = T_1(|f|)$ for every measurable f.

Because the norm on X is absolutely continuous, if $\{g_n\}$ is any sequence in X which converges a.e. to g in X, and if $|g_n| \le |g|$, then $g_n \to g$ in norm. Hence, for a sequence $\{\sigma_n\}$ of simple functions converging a.e. to f, with $|\sigma_n| \le |f|$, we have

$$U\sigma_n \to Uf$$

and

$$U\sigma_n = hT_1\sigma_n \to hT_1f,$$

so that (75) holds for every $f \in X$.

When the measure space is σ-finite, and for $\Omega = \cup\Omega_n$, we proceed as in the proof of Theorem 5.2.11 (or as in the proof of Lamperti's Theorem). For each n, we have from the finite case an isometry U_n induced by U on $L^\varphi(\Omega_n)$ given by $U_n f_n = h_n T'_n f_n$, where T'_n is the operator induced by the set transformation T_n on the measurable subsets of Ω_n. As before, we define $h(t) = h_n(t)$ for $t \in \Omega_n$ and $T(E) = \cup T(E \cap \Omega_n)$. The result now follows with T_1 as the operator induced by T. Again, absolute continuity of the norm is used.

$\square$

In each of the last two theorems, we assumed that the measure was purely nonatomic. In fact, however, it is only in the proof of Theorem 5.2.11 that the nonatomic hypothesis is used. In the case of a general σ-finite measure space (Ω, Σ, μ), the number of atoms must be countable, and so the set Ω_a consisting of the union of all the atoms is in Σ, as is its complement Ω_N. If Σ_a and Σ_N denote the corresponding σ-algebras of measurable sets, then we can write

$$L^\varphi(\Omega, \Sigma, \mu) = L^\varphi(\Omega_a, \Sigma_a, \mu) + L^\varphi(\Omega_N, \Sigma_N, \mu)$$

as a sum of two Orlicz spaces, one of which (written henceforth as ℓ^φ) is over a purely atomic measure and so is a sequence space, and the other (written as L_N^φ) over a purely nonatomic measure.

Suppose $L^\varphi \neq L^2$ and H is a Hermitian operator on L^φ. Let A be a nonzero atom and suppose further that $H\chi_A$ is different from zero on a purely nonatomic set B with $\mu(B) > 0$, so that $\int_B H\chi_A \neq 0$. In exactly the same way as in the proof of Theorem 5.2.11 we get that

$$\varphi'\left(\frac{\alpha}{\|\alpha\chi_B + \chi_A\|}\right) = \alpha\varphi'\left(\frac{1}{\|\alpha\chi_B + \chi_A\|}\right)$$

for all $\alpha > 0$. Since B contains no atoms, we may let $\mu(B) \to 0$ and get

$$\varphi'\left(\frac{\alpha}{\|\chi_A\|}\right) = \alpha\varphi'\left(\frac{1}{\|\chi_A\|}\right)$$

for all α, so that $\varphi(t) = ct^2$ for all $t \geq 0$. However, this contradicts the fact that $L^\varphi \neq L^2$, and we must conclude that $H\chi_A \in \ell^\varphi$.

Now suppose B contains no atoms and $\mu(B) > 0$. Then from (74), we have $\int_A H\chi_B d\mu = 0$ if and only if $\int_B H\chi_A d\mu = 0$ where A is any atom. By what we showed above, $\int_B H\chi_A = 0$, so that $\int_A H\chi_B = 0$. Since this holds for any atom A, we conclude that $H\chi_B \in L_N^\varphi$. It follows from the density of the simple functions, that L_N^φ and ℓ^φ are invariant under H.

Since we know how a Hermitian operator behaves in the purely nonatomic case, it remains to characterize it in the atomic case. We are going to impose the condition that ℓ^φ has a *1-symmetric* basis, that is, a symmetric Schauder basis with symmetric constant 1. This means that ℓ^φ is a sequence space in which the unit vectors e_j (which correspond to the characteristic functions of the atoms) form a basis with the property that $\{e_{\pi(j)}\}$ is a basis equivalent to the original basis for any permutation π of the positive integers. Furthermore, the *symmetric basis constant,*

$$K = \sup_{\theta, \pi} \sup_{\|\sum a(n)e_n\| \leq 1} \| \sum_{n=1}^{\infty} a(n)\theta_n e_{\pi(n)} \|$$

is one, where $\theta = (\theta_n)$ is a sequence of plus and minus ones.

Under the above assumptions, let us denote our Orlicz sequence space ℓ^φ by E. It should be noted that assumption of the δ_2-condition itself implies that E has a symmetric basis. The presence of a 1-symmetric basis actually requires that every operator V of the form

(77) $Vx = (h(j)x(\pi(j)))$

(where we write $y = (y(k))$ in place of $y = \sum_{k=1}^{\infty} y(k)e_k$) is an isometry for any permutation π of the positive integers and any sequence $(h(k))$ of modulus one scalars. It is known that the biorthogonal functionals $\{e_k^*\}$ corresponding to a 1-symmetric basis form a 1-symmetric basis for their closed linear span. In the proof of the next theorem, we will let $\langle x, y^* \rangle$ denote the pairing defined by

$$\langle x, y^* \rangle = \sum x(n)\overline{y(n)}, \quad x \in E, \quad y \in E^*.$$

We note that if $x^* \in E^*$ such that $\|x^*\| = \|x\|$, $\langle x, x^* \rangle = \|x\|^2$, and T is hermitian, then $\langle Tx, x^* \rangle$ is real for every $x \in E$.

5.2.13. THEOREM. *(Arazy, Tam) Let $E = \ell^\varphi$ be an Orlicz sequence space, different from ℓ^2, whose standard unit vectors form a 1-symmetric basis for E. A bounded operator H on E is Hermitian if and only if there is sequence $a = (a(n))$ of real numbers so that $a \in \ell^\infty$ and such that $Hx = (a(n)x(n))$ for all $x \in E$.*

PROOF. The sufficiency of the condition follows from Lemma 5.2.9.

For the necessity, we begin by letting $t_{n,k} = \langle Te_k, e_n^* \rangle$, where T is assumed to be Hermitian. By the remark preceding the statement of the theorem, $a(n) = t_{n,n}$ is real for each n. To prove the theorem, we must show that $t_{n,k} = 0$ for $n \neq k$.

Assume $n \neq k$. We first show that $t_{n,k} = \overline{t_{k,n}}$. For a given θ between 0 and 2π, we let $x(\theta) = e_n + e^{i\theta}e_k$. Then $\|x(\theta)\|_E = \|e_n + e_k\|_E$ and if we take $x^*(\theta) = e_n^* + e^{i\theta}e_k^*$, we must have

$$2 = \langle x(\theta), x^*(\theta) \rangle \leq \|x(\theta)\|_E \|x^*(\theta)\|_{E^*}$$

so that $\|x^*(\theta)\|_{E^*} \geq 2/\|x(\theta)\|_E$. Now suppose that π is the permutation of the positive integers which interchanges n and k and fixes the other indices. For a given $x = \sum x(n)e_n$ let $\pi(x)$ denote the vector obtained as a result of the permutation π on the coordinates. By the symmetry condition, $\|x\| = \|\pi(x)\|$ and the norm of the average of x and $\pi(x)$ is less than or equal to $\|x\|$. From this we get

$$\frac{|x(k) + x(n)|}{2}\|e_n + e_k\| \le \|x\|.$$

Applying the above inequality to elements x of norm 1, we may conclude that $\|x^*(\theta)\|_{E^*} \le 2/\|e_n + e_k\|$. Putting this together with the opposite inequality obtained above, we have $\|x(\theta)\|_E \|x^*(\theta)\|_{E^*} = 2$.

Let $\psi(x(\theta)) = \|x(\theta)\|x^*(\theta)/\|x^*(\theta)\|$. Then $\|\psi(x^*(\theta))\| = \|x(\theta)\|$ and

$$\langle x(\theta), \psi(x^*(\theta)) \rangle = \|x\|^2$$

so that $\langle Tx(\theta), \psi(x^*(\theta)) \rangle$ is real. It is clear from this that $\langle Tx(\theta), x^*(\theta) \rangle$ must also be real, and a straightforward calculation yields

$$t_{n,n} + t_{k,k} + e^{i\theta}t_{n,k} + e^{-i\theta}t_{k,n} \text{ is real.}$$

Therefore, $t_{n,k} = \overline{t_{k,n}}$.

Assume there exist $1 < n < \infty$, $x = \sum_{k=1}^n x(k)e_k$, $y^* = \sum_{k=1}^n y(k)e_k^*$ so that

(i) $x(k) \ge 0, y(k) \ge 0$;

(ii) $\|x\|_E = \|y^*\|_{E^*} = \langle x, y^* \rangle = 1$;

(iii) x and y^* are linearly independent.

Assuming x, y^* as above, and for $\theta = (\theta_1, \theta_2, \ldots, \theta_n)$, $0 \le \theta_k \le 2\pi$ let

$$x(\theta) = \sum_1^n e^{i\theta_k} x(k)e_k,$$

$$y^*(\theta) = \sum_1^n e^{i\theta_k} y(k)e_k^*.$$

Now, $\|x(\theta)\| = \|y^*(\theta)\| = 1\langle x(\theta), y^*(\theta) \rangle$ for all choices of θ and $\langle Tx(\theta), y^*(\theta) \rangle$ must be real for all such θ. As a result,

$$0 = \Im\langle Tx(\theta), y^*(\theta) \rangle$$

$$= \Im \sum_{k,l=1}^n e^{i(\theta_k - \theta_l)} x(k)y(l)t_{l,k}$$

$$= \frac{-i}{2} \sum_{k \ne l} x(k)y(l)(e^{i(\theta_k - \theta_l)}t_{l,k} - e^{i(\theta_l - \theta_k)}t_{k,l})$$

$$= \frac{-i}{2} \sum_{k \ne l} e^{i(\theta_k - \theta_l)}t_{l,k}(x(k)y(l) - x(l)y(k)).$$

This calculation clearly implies that

$$t_{l,k}(x(k)y(l) - x(l)y(k)) = 0 \quad k \ne l.$$

Since x and y^* are not multiples of each other, there exist $k \ne l$ so that $x(k)y(l) \ne x(l)y(k)$ and thus $t_{l,k} = 0$. We can replace x, y^* by $x_\pi = \sum_1^n x(k)e_{\pi(k)}$, $y_\pi^* = \sum_1^n y(k)e_{\pi(k)}^*$ for an appropriate permutation to get the above for any pair k, l. Hence, $t_{k,l} = 0$ for any pair $k \ne l$.

The proof will be complete when we prove the assumption made above. To that end, suppose there exists no pair x, y^* satisfying (i),(ii),(iii). Hence, every nonzero $x \in E$ with finite expansion with respect to $\{e_n\}$ has a unique supporting functional which is proportional to x, denoted by $\alpha(x)x$. It follows that

$$\|x\|_E^2 = \alpha(x)\|x\|_{\ell^2}^2.$$

By taking the Gateaux derivative of the above equation in the direction of x, we can conclude that α is constant on each ray out from the origin, and since $\alpha(e_1) = 1$, we are finally led to the conclusion that $\|x\|_E = \|x\|_{\ell^2}$. This contradicts the hypothesis that $E \neq \ell^2$.   □

We will give a more general form of this result in a later chapter.

The form of the Hermitians given by Theorem 5.2.13 gives rise to the characterization of isometries for the purely atomic case as described by (77). We close this section by stating a theorem characterizing isometries.

5.2.14. THEOREM. *Let $L^\varphi = \ell^\varphi + L_N^\varphi$, where ℓ^φ is as in the statement of Theorem 5.2.13 and $L^\varphi \neq L^2$. If U is an isometry on L^φ, then there is a regular set isomorphism T of the underlying measure space and a fixed function h such that*

$$Uf(\cdot) = h(\cdot)T_1f(\cdot)$$

for all $f \in L^\varphi$, where T_1 is the operator induced by T. Furthermore, U can be written as $U = U_a + U_N$ where U_a and U_N are isometries on ℓ^φ and L_N^φ, respectively.

5.3. Zaidenberg's Generalization

The Orlicz spaces we have been studying are a special case of a more general class of spaces called *rearrangement invariant* or *symmetric* spaces. Zaidenberg has managed to characterize the surjective isometries on this class and we want to give the details of his arguments. Since there are slightly different formulations of the notion of rearrangement invariant given in the literature, we begin by giving a careful statement of the form we intend to use.

5.3.1. DEFINITION. *A closed subspace E of a Banach function space X on a measure space (Ω, Σ, μ) is called an* ideal space *if $f \in E$ and g is a measurable function such that $|g| \leq |f|$, then $g \in E$ and $\|g\|_X \leq \|f\|_X$. (Note that g must be in X by Definition 5.1.2.) If, for $f \in E$ and g a measurable function such that $|f|$ and $|g|$ are equimeasurable, it follows that $g \in E$ and $\|g\| = \|f\|$, we say that the ideal space E is* rearrangement invariant*(r.i.) We let Σ_0 denote the elements of Σ with finite measure, and assume that our (symmetric) space E has the property that $\chi_A \in E$ for every $A \in \Sigma_0$. We also assume that the measure space is σ-finite.*

An important assumption that was prominent in Lumer's results about Orlicz spaces was that the norm was absolutely continuous. In Zaidenberg's work, that assumption has been removed. The following lemma is useful in understanding how we can get around the requirement of absolute continuity of the norm. In the remainder of this section, X will denote a Banach function space on a purely nonatomic measure space (Ω, Σ, μ).

5.3.2. LEMMA. *Suppose E is an ideal space contained in X with $\mu(\Omega) < \infty$. If f is a bounded element of E, then there exists a sequence $\{\sigma_n\}$ of simple functions in E converging almost everywhere to f such that $\|f - \sigma_n\|_X \to 0$.*

PROOF. It is enough to prove this for $f \geq 0$. Since f is bounded and E is an ideal space, there is a sequence $\{\sigma_n\}$ of simple functions in E converging uniformly to f. Letting X' denote the *associate* space for X, it follows from the definition of the associate norm in a Banach function space and [24, Theorem 2.7] that

$$\|f - \sigma_n\|_X = \|f - \sigma_n\|_{X''} = \sup\{\int_\Omega |f - \sigma_n||g|d\mu : g \in X', \|g\|_{X'} \leq 1\}.$$

For each such g,

$$\int_\Omega |f - \sigma_n||g|d\mu \leq \|f - \sigma_n\|_\infty \int_\Omega |g|d\mu$$
$$\leq \|f - \sigma_n\|_\infty C_\Omega$$

where the last inequality follows from Definition 5.1.1(vii). The conclusion of the Lemma is clear. □

Before continuing, we wish to mention some terminology that is becoming widely used in describing isometries. As we saw in Chapter 2, isometries on continuous function spaces are often describable as *weighted composition operators*, and isometries of this type are called *elementary*. In that spirit, we will refer to an isometry of the form given by (75) in the statement of Theorem 5.2.12 as *almost-elementary*. When the regular set isomorphism T which arises in this case can be given by a point transformation, then an almost-elementary isometry is actually elementary. We begin with Zaidenberg's observation that an invertible isometry on an ideal space which has almost-elementary form when acting on the characteristic functions is an almost-elementary operator.

5.3.3. THEOREM. *(Zaidenberg) Let E be an ideal in the space of measurable functions on (Ω, Σ, μ) which contains the characteristic functions χ_A for $A \in \Sigma_0$. Suppose U is an invertible isometry on E, T is a regular set isomorphism on Σ, and h is a measurable function such that*

(78) $$U\chi_A(t) = h(t)\chi_{TA}(t) \quad for \ A \in \Sigma.$$

Then

(79) $$Uf(t) = h(t)T_1 f(t) \quad a.e.,$$

where T_1 is the operator induced by T.

PROOF. Let $f \in E$. Then we may construct a sequence $\{\Omega_n\}$ of sets of finite measure in Σ such that f is bounded on each Ω_n. Using the fact that condition (78) holds for simple functions, Lemma 5.3.2, and the properties of T_1, we may conclude that

$$(80) \qquad U(f\chi_{\Omega_n}) = hT_1(f) \cdot \chi_{T(\Omega_n)}.$$

If we knew that

$$(81) \qquad U(f\chi_{\Omega_n}) = (Uf)\chi_{T(\Omega_n)},$$

then we would have

$$Uf(t) = h(t)T_1 f(t) \text{ a.e. on } T(\Omega_n),$$

and the conclusion of the Theorem would follow. Hence, to complete the proof, it suffices to establish (81), which can be readily seen to be equivalent to

$$(82) \qquad U(f\chi_{(\Omega \setminus \Omega_n)}) \cdot \chi_{T(\Omega_n)} = 0.$$

Hence we will assume that (82) does not hold and consequently, there exists a positive integer n such that $g \cdot \chi_{T(\Omega_n)} \neq 0$, where $g = U(f\chi_{(\Omega \setminus \Omega_n)})$. We observe that $h(t) \neq 0$ a.e. on $\Omega = T(\Omega)$. For if $h = 0$ on a set of positive measure, there would exist a set B of positive measure such that

$$U\chi_B(t) = h(t)\chi_{TB}(t) = 0,$$

contradicting the fact that U is an isometry.

Suppose, then, that there is a measurable set $A \subset T(\Omega_n)$ and constants $\kappa > 0$, $\delta > 0$ such that $0 < \mu(T^{-1}(A)) < \infty$ and

$$(83) \qquad \kappa|h(t)\chi_A(t)| \geq |g(t)\chi_A(t)| \geq \delta|h(t)\chi_A(t)|.$$

Let $\epsilon(t) = T_1^{-1}(\exp(i[\arg(g(t)) - \arg(h(t))])\chi_A(t))$. Then

$$\arg(h(t)T_1\epsilon(t)) = \arg(g(t)) \text{ for } t \in A.$$

If we define f_τ on $\mathbb{R}$ by

$$f_\tau(t) = f(t)\chi_{\Omega \setminus \Omega_n} + \tau\epsilon(t)$$

then we obtain

$$(84) \qquad \|f_\tau\| = \|f_{-\tau}\| = \|g \pm \tau h T_1 \epsilon\| = \|g\chi_{\Omega \setminus A} + (\tau|h| \pm |g|)\chi_A\|.$$

To establish (84), note first that since the support of $\epsilon(t)$ is in Ω_n, we have

$$|f_\tau(t)| = |f(t)\chi_{\Omega \setminus \Omega_n} + \tau\epsilon(t)|$$
$$= |f(t)\chi_{\Omega \setminus \Omega_n} - \tau\epsilon(t)| = |f_{-\tau}|$$

and since E is an ideal,

$$\|f_\tau\| = \|f_{-\tau}\|.$$

Now by the definition of $\epsilon(t)$ we obtain

$$|g \pm \tau h T_1(\epsilon)| = |g\chi_{\Omega\setminus A} + (|g| \pm \tau|h|)\exp(i\arg g)\chi_A|$$
$$= |g\chi_{\Omega\setminus A} + (|g| \pm \tau|h|)\chi_A|$$

where the last equality follows because of the disjointness of supports of the two terms. The equality of norms between the functions on the left and right of the above display then holds by the ideal property. We note here that since $\epsilon(t)$ is a function with modulus 1, we can conclude from (78) and Lemma 5.3.2 that $U(\epsilon) = hT_1\epsilon$. The equality

$$\|f_\tau\| = \|g + \tau h T_1\epsilon\|$$

is clear from the definition of g and the fact that U is an isometry. This completes the verification of (84).

The equalities in (84) show, in particular, that

$$\|g\chi_{\Omega\setminus A} + (\tau|h| + |g|)\chi_A\| = \|g\chi_{\Omega\setminus A} + (\tau|h| - |g|)\chi_A\|$$

and by the inequalities in (83) we can conclude that for $\tau \geq \kappa$,

$$\|g\chi_{\Omega\setminus A} + (\tau - \delta)|h|\chi_A\| = \|g\chi_{\Omega\setminus A} + (\tau + \delta)|h|\chi_A\|.$$

If we let

$$\psi(t) = \|f\chi_{\Omega\setminus A} + \tau|h|\chi_A\|,$$

then ψ is a convex, increasing function which satisfies the identity $\psi(\tau + \delta) = \psi(\tau - \delta)$ for $\tau \geq \kappa$. This would imply that $\psi(t)$ is constant for $t \geq \kappa - \delta$, which is a contradiction.

□

The theorem we have just proved will be an indispensable tool in what follows. Our next step is to obtain a characterization of the Hermitian operators on E. The statement looks much the same as Lumer's Theorem 5.2.11, and says that Hermitian operators are multiplications by real L^∞ functions.

5.3.4. THEOREM. *(Zaidenberg) Let E be an r.i. space such that the norm on E is not proportional to the norm of the space $L^2(\Omega, \Sigma, \mu)$. If H is a bounded Hermitian operator on E, then there exists $h \in L^\infty(\mu)$ such that h is real valued and $Hf = hf$ for all $f \in E$. Furthermore, $\|H\| = \|h\|_\infty$.*

PROOF. It suffices to prove that if H is Hermitian, then

(85) $$\chi_{\Omega\setminus A} \cdot H\chi_A = 0 \text{ for all } A \in \Sigma_0.$$

Suppose (85) holds. It follows easily that

(86) $$\chi_A H\chi_B = \chi_B H\chi_A \text{ for all } A, B \in \Sigma_0,$$

and so there exists a measurable function h such that

(87) $$H\chi_A = h\chi_A \text{ for every } A \in \Sigma_0.$$

(The function h would be given by $H(\chi_\Omega)$ in the case of finite measure, with the appropriate adjustments made in the σ-finite case.)

If the function h were not essentially bounded, then we could choose, for each positive integer N, an $A_N \in \Sigma_0$ with $\mu(A_N) > 0$ such that $|h| \geq N$ on A_N. Thus,

$$N\|\chi_{A_N}\| \leq \|h\chi_{A_N}\| = \|H\chi_{A_N}\| \leq \|H\|\|\chi_{A_N}\|$$

which would contradict the boundedness of H. We conclude that

$$\|h\|_\infty \leq \|H\|.$$

It is clear from (87) that $H(g\chi_A) = gH(\chi_A)$ for any simple function g, and by arguments similar to those of the previous theorem, using Lemma 5.3.2, we can show that

$$H(g\chi_A) = gH(\chi_A)$$

for any g which is essentially bounded. In particular, we have, by induction, that $H^n(\chi_A) = h^n\chi_A$ for each n, and hence,

$$(88) \qquad e^{itH}(\chi_A) = e^{ith}\chi_A$$

for every $A \in \Sigma_0$. From Theorem 5.3.3 we must have

$$e^{itH}(f) = e^{ith}(f)$$

for all $f \in E$. By straightforward calculation, using the series expansions, we get that

$$\frac{d}{dt}e^{itH}(f) = iH(f)$$

and

$$\frac{d}{dt}e^{ith}(f) = ihf.$$

Therefore, $H(f) = hf$ and it is immediate that $\|H\| \leq \|h\|_\infty$. Coupled with our earlier observation, this gives $\|H\| = \|h\|_\infty$. To complete this segment of the argument, we must show that h is real valued.

For each $f \in E$, we can write $f(t) = \exp(i\theta(t))f_r(t)$, where f_r is real valued and θ is a function on Ω with values in $[0, 2\pi]$. Let E_r denote the space of all real valued functions in E with norm inherited from E. Then $f_r \in E_r$ and there exists f_r^* in E_r^* such that

$$f_r^*(f_r) = \|f_r\|^2, \quad \text{and} \quad \|f_r^*\| = \|f_r\|.$$

Let f_0^* denote the bounded linear functional on E defined by

$$f_0^*(g_1 + ig_2) = f_r^*(g_1) + if_r^*(g_2).$$

Note that $\|f\| = \|f_r\|$ and it can be shown that $\|f_0^*\| = \|f_r^*\|$. If we define f^* by $f^*(g) = f_0^*(\exp(-i\theta(\cdot))g)$, then f^* has the necessary properties for $[g, f] = f^*(g)$ to define a semi-innerproduct on E which is compatible with its norm. It follows that

$$[Hf, f] = f_r^*(\Re(h)f_r) + if_r^*(\Im(h)f_r),$$

and since $[Hf, f]$ must be real for every $f \in E$, we have that $f_r^*(\Im(h)f_r)$ is real for every such f. We conclude that $\Im(h) = 0$.

The conclusion of the theorem holds then as long as (85) is satisfied. Our task now is to show that a denial of (85) leads to a contradiction. Hence suppose that there is a Hermitian operator H_0 and a set $A_0 \in \Sigma_0$ such that

$$f = \chi_{\Omega \setminus A_0} \cdot H_0 \chi_{A_0} \neq 0.$$

The function f is the limit of a sequence of simple functions, and by Egoroff's theorem, it is the uniform limit of such a sequence on a set of positive measure, we may find a measurable subset A_1 of $\Omega \setminus A_0$ and a nonzero scalar λ such that

$$(89) \qquad \|\chi_{A_1} H_0 \chi_{A_0} - \lambda \chi_{A_1}\| < |\lambda| \|\chi_{A_1}\|.$$

Since $\lambda \neq 0$, we must have $\chi_{A_1} H_0 \chi_{A_0} \neq 0$. Let $A_2, \ldots, A_n$ be any collection of measurable sets in $\Omega \setminus A_0$ which are disjoint from each other and from A_1. Assume also that $\mu(A_j) = \mu(A_1)$ for each $j = 2, 3, \ldots, n$ and let E^{n+1} be the subspace of E spanned by $\chi_{A_0}, \chi_{A_1}, \ldots, \chi_{A_n}$. There is a projection P on E^{n+1} which commutes with each projection P_{A_j} (where $P_{A_j}(f) = \chi_{A_j} f$). Thus $P \chi_{A_j} = \chi_{A_j}$ for each $j = 0, 1, \ldots, n$ and if we let $\tilde{H}_0 = P H_0 P$, then $\tilde{H}_0$ is Hermitian because it can be shown that $\|I - it P H_0 P\| = 1 + o(t)$. Furthermore, E^{n+1} is invariant under $\tilde{H}_0$ and

$$\|\chi_{A_1}(\tilde{H}_0 \chi_{A_0}) - \lambda \chi_{A_1}\| = \|P(\chi_{A_1}(H_0 \chi_{A_0}) - \lambda \chi_{A_1}\|$$
$$\leq |\lambda| \|\chi_{A_1}\|$$

by (89) and properties of P. As before, we see from the above inequality that

$$\chi_{A_1} \tilde{H}_0 \chi_{A_0} \neq 0.$$

The sets $\{A_j\}_1^n$ all have the same measure; therefore, their characteristic functions are equimeasurable and so have the same norms. If $k \neq j$, and λ is any scalar, then

$$\|\chi_{A_j}\| \leq \|\chi_{A_j} + \lambda \chi_{A_k}\|$$

and we conclude that the characteristic functions are pairwise orthogonal in the sense of James as well as in the usual sense in case the norm is given by an inner product. Consider the operator U defined on E^{n+1} by

$$(90) \qquad U(c_1 \chi_{A_1} + \cdots + c_n \chi_{A_n} + c_0 \chi_{A_0}) = \sum_{j=1}^{n} \epsilon_j c_j \chi_{A_{\pi(j)}} + c_0 \chi_{A_0}$$

where $|\epsilon_j| = 1$ and π is a permutation of the indices $1, 2, \ldots, n$. If g is an element of E^{n+1} and given as in the parentheses on the left above, and if $f = Ug$, then f and g are equimeasurable. Since E is an r.i. space, we conclude that $\|Uf\| = \|f\|$ and U is an isometry. By a result of Rolewicz [**268**, Lemma IX.8.4], if the group G of isometries on E^{n+1} contains all operators of the form given in (90), then G contains all orthogonal transformations on the space Y generated by $\chi_{A_1}, \ldots, \chi_{A_n}$, or G consists entirely of operators of the form

(91) $$U\left(\sum_{j=0}^{n} t_j \chi_{A_j}\right) = \sum_{j=0}^{n} \epsilon_j t_j \chi_{A_{\pi(j)}}$$

where $|\epsilon_j| = 1$ and π is a permutation of $\{0, 1, \ldots, n\}$.

Since $\tilde{H}_0$ is a Hermitian operator for which E^{n+1} is invariant, the restriction of $\exp(it\tilde{H}_0)$ is an isometry for every real t. If such an isometry is of the form (91), then we must have

(92) $$e^{it\tilde{H}_0}\chi_{A_0} = \epsilon_0 \chi_{A_j}$$

where $j = \pi(0)$. If we suppose that $j \neq 1$, then upon expanding the left side of (92) and multiplying both sides by χ_{A_1} we obtain an equality of the form

$$it(f + tg) = 0,$$

where $f, g \in E$ and $f \neq 0$. Since this must hold for all t, we have a contradiction. However, if $j = 1$, then it follows from (92) that

$$\|\chi_{A_0}\| = \|e^{it\tilde{H}_0}\chi_{A_0} = \|\epsilon_0\chi_{A_1}\| = \|\chi_{A_1}\|.$$

This cannot be since the set A_0 was fixed and the set A_1 could have been chosen to have a measure that is positive but as small as we wish. We conclude then, by the aforementioned result of Rolewicz, that the group G consists entirely of orthogonal transformations. Consequently, the space Y is a Hilbert space with orthogonal basis $\{\chi_{A_1}, \ldots, \chi_{A_n}\}$. Since these basis elements all have the same norm, we must have, for each $y \in Y$, that

(93) $$\|y\|^2 = \|\sum_{j=0}^{n} c_j \chi_{A_j}\|^2 = \|\chi_{A_1}\|^2 \sum_{j=1}^{n} |c_j|^2.$$

Suppose that A and B are disjoint elements of Σ_0. For n sufficiently large, we may choose sets $A_1^{(n)}, \ldots, A_{p_n}^{(n)}$ and $B_1^{(n)}, \ldots, B_{q_n}^{(n)}$ which are pairwise disjoint, all of the same measure and such that $\bigcup_j A_j^{(n)} \subset A$, $\bigcup_j B_j^{(n)} \subset B$. Suppose also that the sets are chosen so that

$$\mu\left(A\backslash\bigcup_j A_j^{(n)}\right) < \frac{\mu(A) + \mu(B)}{2^n}$$

and the same inequality holds with the A's replaced by B's. If N is sufficiently large for this, we make such selections for every $n > N$ and in such a way that

$$\bigcup_j A_j^{(n)} \subset \bigcup_j A_j^{(n+1)} \subset A$$

and

$$\bigcup_j B_j^{(n)} \subset \bigcup_j B_j^{(n+1)} \subset B.$$

For positive numbers α and β,

$$\sum_{j=1}^{p_n} \alpha \chi_{A_j^{(n)}} + \sum_{j=1}^{q_n} \beta \chi_{B_j^{(n)}} \nearrow \alpha \chi_A + \beta \chi_B.$$

By the Fatou property of Banach function spaces, it is therefore true that

$$\| \sum \alpha \chi_{A_j^{(n)}} + \sum \beta \chi_{B_j^{(n)}} \| \nearrow \| \alpha \chi_A + \beta \chi_B \|.$$

If the sets A, B are both in $\Omega \backslash \Sigma_0$, then the sets $A_j^{(n)}, B_j^{(n)}$ satisfy the requirements which led to (93). Hence,

$$\| \sum_{j=1}^{p_n} \alpha \chi_{A_j^{(n)}} + \sum_{j=1}^{q_n} \beta \chi_{B_j^{(n)}} \|^2 = \| \chi_{A_1^{(n)}} \|^2 (p_n |\alpha|^2 + q_n |\beta|^2)$$

$$= \| \sum \alpha \chi_{A_j^{(n)}} \|^2 + \| \sum \beta \chi_{B_j^{(n)}} \|^2$$

$$\nearrow \| \alpha \chi_A \|^2 + \| \beta \chi_B \|^2$$

$$= |\alpha|^2 \| \chi_A \|^2 + |\beta|^2 \| \chi_b \|^2.$$

Thus we have

(94) $$\| \alpha \chi_A + \beta \chi_B \|^2 = |\alpha|^2 \| \chi_A \|^2 + |\beta|^2 \| \chi_B \|^2.$$

The properties of an ideal space guarantee that the above equality holds for any scalars α, β. If either of the sets A or B (or both) lie inside A_0, then we can choose sets $\tilde{A}_j^{(n)}$ and $\tilde{B}_j^{(n)}$ (if necessary) in $\Omega \backslash A_0$ with the same measures as the $A's$ and $B's$ originally chosen as in the argument above, so that the function $\sum \alpha \chi_{\tilde{A}_j^{(n)}} + \sum \beta \chi_{\tilde{B}_j^{(n)}}$ will be equimeasurable with the original function. From this we can conclude that (94) holds for any two disjoint sets A, B in Σ_0.

It is straightforward to show, using (94), that for any two simple functions σ and τ,

$$\| \sigma + \tau \|^2 + \| \sigma - \tau \|^2 = 2 \| \sigma \|^2 + 2 \| \tau \|^2.$$

Let E^b denote the closure of the set of simple functions with support in Σ_0. For any $f, g \in E^b$, we can take limits of simple functions to show that

$$\| f + g \|^2 + \| f - g \|^2 = 2 \| f \|^2 + 2 \| g \|^2.$$

Therefore, E^b is a Hilbert space and since $(E^b)' = E'$, we get that $E^b = E$ and E is a Hilbert space.

In fact, we can show a bit more. Let B be a given measurable set with $\mu(B) = 1$, and let $A \in \Sigma$. Then by an argument involving the partitioning of the sets A and B into collections of sets of equal measure and (93), we obtain

$$\| \chi_A \|^2 = \mu(A) \| \chi_B \|^2 = \left(\int \chi_A d\mu \right) \| \chi_B \|^2.$$

This equality can be extended to simple functions and by limits to all elements of E so that we have

$$\| f \|^2 = \| \chi_B \|^2 \| f \|_{L^2(\mu)}^2.$$

This statement contradicts the assumption in the theorem about E and completes the proof. $\square$

Now that we have cleared away the rather formidable underbrush, the path to Zaidenberg's main theorem, the characterization of isometries on r.i. spaces, becomes quite clear. We remind the reader that throughout this section, we are assuming that the measure spaces are purely nonatomic.

5.3.5. THEOREM. *(Zaidenberg) Let E_1 and E_2 be r.i. spaces associated with measure spaces*
$(\Omega_1, \Sigma_1, \mu_1)$, $(\Omega_2, \Sigma_2, \mu_2)$, respectively, and assume that the norm on E_1 is not proportional to the norm on $L^2(\mu_1)$. If U is a surjective isometry from E_1 onto E_2, then U is almost elementary; that is, there exist a measurable function h and a regular set isomorphism T from Σ_1 onto Σ_2 such that

$$Uf(t) = h(t)T_1f(t) \quad for \ all \ f \in E_1,$$

where as before, T_1 is the operator induced by T.

PROOF. Since U is an isometry, the norm in E_2 is not proportional to an L^2-norm either. By the previous theorem, each Hermitian operator is given by multiplication by a real-valued L^∞-function. Furthermore, the mapping $H \to UHU^{-1}$ sets up an algebraic isomorphism between the Hermitian operators on E_1 and the Hermitian operators on E_2. From this we can construct a regular set isomorphism T in exactly the same way as it was done in the proof of Lumer's Theorem 5.2.12 and obtain a measurable function h so that equation (78) of Theorem 5.3.3 is satisfied.

We will comment on the selection of the function h since in the proof of Lumer's Theorem, we used the absolute continuity of the norm. If the measure of Ω_1 is finite, then we can take $h = U(1)$ as usual. Otherwise, we have

$$\Omega_1 = \bigcup_{n=1}^{\infty} \Omega_n^{(1)}$$

where $\mu_1(\Omega_n^{(1)}) < \infty$ for each n, and we let

$$h_n = U(\chi_{\Omega_n^{(1)}}).$$

For $t \in T(\Omega_n^{(1)})$, we have $h_n(t) = h_{n+1}(t)$ and for any $A \in \Sigma_1$ with finite measure,

$$U(\chi_A \chi_{\Omega_n^{(1)}})(t) = h(t)\chi_{TA}(t).$$

By the argument as given in the proof of Theorem 5.3.3, since χ_A is bounded on each $\Omega_n^{(1)}$, we have

$$U(\chi_A)\chi_{T(\Omega_n^{(1)})} = U(\chi_A \chi_{\Omega_n^{(1)}}).$$

From this we get

$$U\chi_A(t) = h(t)\chi_{TA}(t),$$

and an application of Theorem 5.3.3 completes the proof. □

5.4. Musielak-Orlicz Spaces

Another way to obtain a generalization of the L^p-spaces is to allow the constant p to be replaced by a measurable function $p(t)$. This results in the Nakano $L^{p(t)}$-spaces which are defined in terms of a modular, I_Φ, similar to the way in which Orlicz spaces are defined. Thus for a Young's-type function $\varphi(u, t) = u^{p(t)}$, and $I_\varphi(f) = \int \Phi(|f(t)|, t)dt$, we say that $f \in L^{p(t)}$ if for some $r > 0$, $I_\Phi(rf) < \infty$, and the norm is given by

$$(95) \qquad \|f\| = \inf\{\epsilon > 0 : I_\Phi(\frac{1}{\epsilon}|f|) \le 1\}.$$

These spaces are neither Orlicz spaces nor rearrangement invariant. (For example, take the basic measure space to be Lebesgue measure on $[0, 1]$ and let $p(t)$ be 3 on $[0, 1/2]$ and 4 on $(1/2, 1]$. The functions $\chi_{[0,1/2]}, \chi_{[1/2,1]}$ are equimeasurable, but have different norms.) They are special cases of a class of spaces called *generalized Orlicz spaces* or *Musielak-Orlicz spaces* and the isometries on this class are known. As in the case of Orlicz spaces, there are two equivalent norms that can be assigned, and we want to show that the isometries are the same for both norms. In the entire section we will assume that we are working with a measure space (Ω, Σ, μ) that is σ-finite, atomless, and separable.

5.4.1. DEFINITION. *A function $\Phi(u, t)$ from $\mathbb{R}_+ \times \Omega$ to $\mathbb{R}_+$ is said to be a Young's function with parameter, or Musielak-Orlicz function, if it is convex and increasing with respect to u, measurable as a function of t, and $\Phi(0, t) = 0$ a.e. In addition, we will assume that*

$$(96) \qquad \lim_{u \to \infty} \frac{\Phi(u, t)}{u} = \infty \quad and \quad \lim_{u \to 0} \frac{\Phi(u, t)}{u} = 0,$$

for almost all $t \in \Omega$.

Given an M-O function Φ (we will use the abbreviation M-O for Musielak-Orlicz), the conjugate function $\Phi^*(u, t)$ defined for nonnegative u and $t \in \Omega$ by

$$(97) \qquad \Phi^*(u, t) = \sup_{v \ge 0}\{uv - \Phi(v, t)\}$$

is also an M-O function satisfying the conditions given in (96). We will have occasion to distinguish between the left and right hand derivatives of $\Phi(u, t)$ with respect to u which will be denoted by $\Phi'_-(u, t)$ and $\Phi'_+(u, t)$, respectively. The subscripts +,- will be omitted when distinctions are unnecessary. We should note also that the conditions given imply that $\Phi^*(u, t) < \infty$ and

$$(98) \qquad \lim_{u \to \infty} \Phi'(u, t) = \infty \quad and \quad \lim_{u \to 0} \Phi'(u, t) = 0.$$

The reader will have noticed the similarity of the discussion above to that given for Orlicz functions in Section 2. Indeed, the M-O functions reduce to the Orlicz functions when $\Phi(u, t)$ is constant as a function of t for each u.

We have adopted the convention of using Φ for an M-O function instead of the φ used for an Orlicz function. We wish to note that the properties given in Theorem 5.2.4(viii) have exact analogues which we will not state explicitly here but which will be needed later. There is one more relationship concerning an M-O function and its conjugate which we will also need.

$$(99) \quad \Phi^*(\lambda\Phi'_-(u,t) + (1-\lambda)\Phi'_+(u,t),t)$$
$$= \lambda\Phi^*(\Phi'_-(u,t),t) + (1-\lambda)\Phi^*(\Phi'_+(u,t),t)$$

for $\lambda \in [0,1]$.

We will let L^Φ denote the set of all measurable, scalar-valued functions f such that

$$I_\Phi(rf) = \int_\Omega \Phi(r|f(t)|,t)d\mu < \infty$$

for some $r > 0$, and E^Φ will denote the set of all $f \in L^\Phi$ such that $I_\Phi(rf) < \infty$ for all $r > 0$. Then E^Φ is a subspace of L^Φ which is equal to L^Φ whenever Φ satisfies a Δ_2-condition (see Definition 5.2.2). The equations (63) and (64) from Section 2 (with φ replaced by Φ) define Banach function norms for which L^Φ is a Banach function space. As in the case of Orlicz space, these two norms are equivalent and it is our goal in this section to show that the groups of isometries corresponding to these norms are the same.

We will begin by characterizing the isometries of L^Φ with respect to the Luxemburg norm

$$\|f\|_l = \inf\{\epsilon > 0 : I_\Phi(f/\epsilon) \leq 1\}.$$

We will show that any such isometry is almost elementary. The approach will be similar to that of Lumer for Orlicz spaces although different techniques are required in some places. In order to minimize the details which must be given, we will operate in a setting that is not the most general possible. It should be noted, however, that we do not assume that $E^\Phi = L^\Phi$ as we did in Section 2.

It is clear from the definition that if $f \in L^\Phi$ and $\|f\| = 1$, then $I_\Phi(f) \leq 1$. Examination of the proof of Lemma 5.2.7 shows that if in addition, $f \in E^\Phi$, then $I_\Phi(f) = 1$. On the other hand, $I_\Phi(f) = 1$ implies that $\|f\| = 1$ for any $f \in L^\Phi$. We state a series of lemmas whose proofs require only minor modifications of those given for similar results in Section 2.

5.4.2. LEMMA. *For any $g \in E^\Phi$, the functional*

$$F_g(f) = C(g)\int_\Omega f(t)sgn(g(t)\Phi'\left(\frac{|g(t)|}{\|g\|},t\right)d\mu$$

where $f \in L^\Phi$ and

$$C(g) = \|g\|^2 / \int_\Omega |g(t)|\Phi'\left(\frac{|g(t)|}{\|g\|},t\right)d\mu$$

is a support functional of g.

5.4.3. LEMMA. *If* $h \in L^\infty(\Omega, \Sigma, \mu)$ *is real valued, the operator* H_h *defined by* $H_h(f) = hf$ *is a bounded Hermitian operator on* L^Φ *and* $\|H_h\| = \|h\|_\infty$.

PROOF. For this we appeal to a part of the argument given in the proof of Theorem 5.3.4. Given $f \in L^\Phi$, construct f^* as in that proof. Then there is a s.i.p. $[,]$ on L^Φ such that $[g, f] = f^*(g)$ and using its definition, it is easy to see that $[Hf, f]$ is real. $\square$

5.4.4. LEMMA. *Let* H *be a Hermitian operator on* L^Φ *and suppose* f, g *are elements in* E^Φ *with disjoint supports* A, B, *respectively. Then*

(100)
$$\int_A Hg(t)\Phi'\left(\frac{|f(t)|}{\|f+g\|}, t\right) sgn f(t) d\mu = \int_B \overline{Hf(t)}\Phi'\left(\frac{|g(t)|}{\|f+g\|}, t\right) \overline{sgn g(t)} d\mu.$$

5.4.5. COROLLARY. *For* A, B *which are disjoint and such that* χ_A, χ_B *are in* E^Φ, *and for any* $\alpha, \beta > 0$ *with* $\|\alpha\chi_A + \beta\chi_B\| = 1$ *we have*

$$\frac{1}{\alpha}\int_A H(\chi_B)(t)\Phi'(\alpha, t) d\mu = \frac{1}{\beta}\int_B \overline{H(\chi_A)(t)}\Phi'(\beta, t) d\mu.$$

From this point on we will assume that for almost all $t \in \Omega$, the function $u \to \frac{\Phi'(u,t)}{u}$ is monotone. Furthermore, we will assume that $\chi_\Omega \in E^\Phi$. This means that any simple function is in E^Φ. The reader should consult the notes at the end of the chapter for a discussion of more general results. We now state and prove a rather technical lemma.

5.4.6. LEMMA. *Let* $A \in \Sigma$ *with* $\mu(A) > 0$. *Suppose* $u \to \frac{\Phi'(u,t)}{u}$ *is nondecreasing and nonconstant for all* $t \in A$. *Then there exists a sequence* $\{\alpha_n\}$ *of positive numbers and a sequence* $\{A_n\}$ *of pairwise disjoint sets such that* $\cup A_n = A$, $\chi_{A_n} \in E^\Phi$, $I_\Phi(\alpha_n \chi_{A_n}) < 1$ *and for all* $\gamma > \alpha_n$, *there is a subset* B *of* A_n *such that*

$$\frac{\Phi'(\gamma, t)}{\gamma} > \frac{\Phi'(\alpha_n, t)}{\alpha_n}$$

for all $t \in B$. *(If we assume that* $\frac{\Phi'(u,t)}{u}$ *is nonincreasing, then the inequality above is reversed.)*

PROOF. We give the proof only for the nondecreasing case.
If for each pair α, β of positive rational numbers we let

$$C_{\alpha\beta} = \left\{ t \in A : \frac{\Phi'(\alpha, t)}{\alpha} < \frac{\Phi'(\beta, t)}{\beta} \right\}$$

then the hypotheses imply that $\cup C_{\alpha\beta} = A$. This family of sets is countable and we relabel them as a disjoint collection $\{C_n\}$. By partitioning the sets C_m if necessary, and relabeling again, we have a disjoint family $\{A_n\}$ whose union is A and for each A_n there are positive rationals α, β such that for each $t \in A_n$,

$$\frac{\Phi'(\alpha, t)}{\alpha} < \frac{\Phi'(\beta, t)}{\beta} \quad \text{and} \quad I_\Phi(\beta\chi_{A_n}) < 1.$$

For each positive rational γ, let

$$G_\gamma = \left\{ t \in A : \frac{\Phi'(\gamma, t)}{\gamma} = \frac{\Phi'(\alpha, t)}{\alpha} \right\}.$$

Then for $t \in A_n$ define the measurable function $\alpha(t)$ by

$$\alpha(t) = \sup \left\{ \gamma \geq \alpha : \frac{\Phi'(\gamma, t)}{\gamma} = \frac{\Phi'(\alpha, t)}{\alpha} \right\} = \sup_{\gamma \geq \alpha} \{ \gamma \chi_{G_\gamma}(t) + \alpha \chi_{A_n \setminus G_\gamma}(t) \}.$$

The left continuity of $u \to \Phi'(u, t)$ implies that

$$\frac{\Phi'(\alpha(t), t)}{\alpha(t)} = \frac{\Phi'(\alpha, t)}{\alpha}$$

and we also have

$$\frac{\Phi'(\gamma, t)}{\gamma} > \frac{\Phi'(\alpha(t), t)}{\alpha(t)}$$

for any $\gamma > \alpha(t)$. If we let

$$\alpha_n = ess \inf_{t \in A_n} \alpha(t),$$

then $\alpha \leq \alpha_n \leq \alpha(t) < \beta$ for $t \in A_n$. Given $\gamma > \alpha_n$ there exists $B \subset A_n$ with positive measure such that $\gamma > \alpha(t)$ for $t \in B$. Then $I_\Phi(\alpha_n \chi_{A_n}) \leq I_\Phi(\beta \chi_{A_n}) < 1$ and the set B satisfies the desired properties. $\qquad \square$

We now take the first step in showing that a Hermitian operator is a multiplier. Let

$$\Omega_0 = \left\{ t \in \Omega : u \to \frac{\Phi'(u, t)}{u} \text{ is constant} \right\}.$$

5.4.7. LEMMA. *Suppose $u \to \frac{\Phi'(u,t)}{u}$ is monotone for any $t \in \Omega$. If H is a Hermitian operator on L^Φ and $B \in \Sigma$, then*

(i) *$supp H(\chi_B) \subset \Omega_0 \cup B$.*
(ii) *If $B \cap \Omega_0 = \emptyset$ then $supp H(\chi_B) \subset B$.*

PROOF. Recall that we are assuming that any simple function is in E^Φ.
(i) Let

$$P = \left\{ t \in \Omega \backslash (\Omega_0 \cup B) : u \to \frac{\Phi'(u, t)}{u} \text{ is nondecreasing} \right\},$$

$$S = \{ t \in \Omega \backslash (\Omega_0 \cup B) : \Re H(\chi_B)(t) \geq 0 \}.$$

We want to show that $\Re H(\chi_B)(t) = 0$ a.e in $P \cap S$. Suppose there is a subset A of $P \cap S$ of positive measure such that $\Re H(\chi_B)$ is positive on A. By Lemma 5.4.6 there is a sequence $\{\alpha_n\}$ of positive numbers and a sequence $\{A_n\}$ of pairwise disjoint sets whose union is A so that for each n, $I_\Phi(\alpha_n \chi_{A_n}) < 1$. Now our assumptions about Φ and Φ' guarantee that $I_\Phi(\gamma \chi_B) \to \infty$ as $\gamma \to \infty$. Hence we may choose $\gamma > 0$ and $\beta_n > \alpha_n$ so that

$$I_\Phi(\alpha_n \chi_{A_n}) + I_\Phi(\gamma \chi_B) = 1$$

and

$$I_\Phi(\beta_n \chi_{A_n}) = 2I_\Phi(\alpha_n \chi_{A_n}).$$

Furthermore, there exist disjoint sets A_{n1}, A_{n2} whose union is A_n for which

$$I_\Phi(\beta_n \chi_{A_{n1}}) = I_\Phi(\beta_n \chi_{A_{n2}} = I_\Phi(\alpha_n \chi_{A_n}).$$

It now follows that

$$I_\Phi(\beta_n \chi_{A_{nj}}) + I_\Phi(\gamma \chi_B) = 1$$

for $j = 1, 2$ and therefore

$$\|\alpha_n \chi_{A_n} + \gamma \chi_B\| = \|\beta_n \chi_{A_{nj}} + \gamma \chi_B\| = 1$$

for $j = 1, 2$. The application of Corollary 5.4.5 leads to

(101)
$$\int_{A_n} H(\chi_B)(t) \frac{\Phi'(\alpha_n, t)}{\alpha_n} d\mu = \int_B \overline{H(\chi_{A_n})(t)} \frac{\Phi'(\gamma, t)}{\gamma} d\mu.$$

Similarly, we get for $j = 1, 2$,

$$\int_{A_{nj}} H(\chi_B)(t) \frac{\Phi'(\beta_n, t)}{\beta_n} d\mu = \int_B \overline{H(\chi_{A_{nj}}(t)} \frac{\Phi'(\gamma, t)}{\gamma} d\mu,$$

and upon adding these two equations we obtain

(102)
$$\int_{A_n} H(\chi_B)(t) \frac{\Phi'(\beta_n, t)}{\beta_n} d\mu = \int_B \overline{H(\chi_{A_n})(t)} \frac{\Phi'(\gamma, t)}{\gamma} d\mu.$$

The combination of (101) and (102) along with the taking of real parts gives

(103)
$$\int_{A_n} \Re H(\chi_B)(t) \left[\frac{\Phi'(\beta_n, t)}{\beta_n} - \frac{\Phi'(\alpha_n, t)}{\alpha_n} \right] d\mu = 0.$$

It is a consequence of Lemma 5.4.6 that corresponding to $\beta_n > \alpha_n$ there is a subset C of $A_n \subset A$ such that

$$\frac{\Phi'(\beta_n, t)}{\beta_n} > \frac{\Phi'(\alpha_n, t)}{\alpha_n}$$

for all $t \in C$. Since equation (103) holds with A_n replaced by C, it then follows that $\Re H(\chi_B)(t) = 0$ a.e. on C which is a contradiction to the choice of A. We conclude that $\Re H(\chi_B) = 0$ a.e. on $P \cap S$. Clearly, similar arguments can be given for the other relevant cases and we must have $H\chi_B$ is zero on all of $\Omega \backslash (\Omega_0 \cup B)$, which finishes the proof of (i).

(ii) We assume that $\Omega_0 \cap B = \emptyset$ and let W denote the set of t in Ω_0 for which the real part of $H\chi_B$ is nonnegative. Then we assume that $\mu(W) > 0$ and observe that W and B are disjoint. We choose α, β such that

$$I_\Phi(\alpha \chi_W) + I_\Phi(\beta \chi_B) = \|\alpha \chi_W + \beta \chi_B\| = 1.$$

By part (i), $\text{supp} H(\chi_W) \subset (\Omega_0 \cup W) \subset \Omega_0$, and upon application of Corollary 5.4.5 we must have

$$\int_W \Re H \chi_B(t) \frac{\Phi'(\alpha, t)}{\alpha} d\mu = 0.$$

However, $\Re H\chi_B \geq 0$ on W and $\Phi'(\alpha,t)/\alpha$ is constant there, so that $\Re H\chi_B = 0$ a.e. on W. This essentially finishes the argument, since the cases for $\Re H\chi_B \leq 0$ and $\Im H\chi_B$ can be treated in the same way. □

We are now ready to characterize the Hermitian operators on L^Φ, the analogue of Theorems 5.2.11 and 5.3.4.

5.4.8. THEOREM. *Suppose $\mu(\Omega_0) = 0$ and $u \to \frac{\Phi'(u,t)}{u}$ is monotone for each $t \in \Omega$. Any bounded linear operator H on L^Φ is Hermitian if and only if*

(104) $$Hf = hf$$

for all $f \in L^\Phi$, where h is a real-valued L^∞ function. In this case, $\|H\| = \|h\|_\infty$.

PROOF. The sufficiency is given by Lemma 5.4.3.

For the necessity, assume H is Hermitian. If $\mu(\Omega) < \infty$, let $h(t) = H\chi_\Omega(t)$. If $\mu(\Omega) = \infty$ and $\Omega = \cup\Omega_n$ where the Ω_n's are disjoint and each of finite measure, let $h(t) = \sum H\chi_{\Omega_n}(t)$. Suppose $A \in \Sigma$ and $A \subset \Omega_n$. Utilizing Lemma 5.4.7 and the disjointness of the Ω_n we see that

$$H(\chi_{\Omega_n})\chi_A = H(\chi_A)\chi_A + H(\chi_{(\Omega_n \setminus A)})\chi_A = H(\chi_A),$$

and

$$H(\chi_A) = h\chi_A.$$

It follows that

$$e^{i\alpha H}(\chi_A) = e^{i\alpha h}\chi_A$$

for all real α, and all $A \subset \Omega_n$. Now $e^{i\alpha H}$ is an isometry since H is Hermitian, and from Theorem 5.3.3 we get that

$$e^{i\alpha H}(f) = e^{i\alpha h}f$$

for every f in L^Φ. From this we conclude that (104) holds exactly as in the proof of Theorem 5.3.4. The fact that h must be a real L^∞ function follows as in the proof of Lemma 5.2.9. □

We are now at the point where we can state and prove the theorem characterizing the isometries on M-O spaces. The proof that such an isometry is almost elementary, that is, of the form $Uf = hT_1f$, is essentially the same as given for Theorem 5.3.5. Recall that if T is a regular set isomorphism on Σ, then $\nu = \mu \circ T^{-1}$ is a measure which is absolutely continuous with respect to μ and so has a Radon-Nikodym derivative with respect to μ which we will denote here by g. Note that g is measurable and that the set of t for which $g(t) = 0$ must be of measure zero by properties of T. We are going to obtain a condition on the weight function h which relates it to the M-O function Φ and the Radon-Nikodym derivative g. This condition, a generalization of the one in Lamperti's Theorem (3.2.5 in Chapter 3), also gives a sufficient condition for the operator $Uf = hT_1f$ to be an isometry. All of this adds considerably to the length of the proof for our main theorem, and before we begin that, it will be convenient to have the following interesting lemma.

5.4.9. LEMMA. *Let ν be a positive measure on Σ with the property that $\nu(A) = 0$ if and only if $\mu(A) = 0$, and let c be a positive number. If $c < \nu(\Omega) < \infty$ and f is a measurable, real function such that $\int_A f(t)d\mu = 0$ for any A with $\nu(A) = c$, then $f = 0$ a.e.*

PROOF. Note first that since μ is atomless, the same is true of ν. Let

$$B_1 = \{t \in \Omega : f(t) > 0\}, \quad B_2 = \{t \in \Omega : f(t) < 0\}.$$

Suppose first that $\mathrm{supp} f$ is all of Ω. Let A be such that $\nu(A) = c$ so that $\int_A f(t)d\mu = 0$. Therefore,

$$\int_{A \cap B_1} f(t)d\mu = -\int_{A \cap B_2} f(t)d\mu.$$

Hence both $\mu(A \cap B_1)$ and $\mu(A \cap B_2)$ are positive, and since $\nu(\Omega \backslash A) > 0$, either $\nu(B_1 \backslash A) > 0$ or $\nu(B_2 \backslash A) > 0$. If, to be specific, we suppose it is the former, then we may choose $D \subset A \cap B_2$ such that $0 < \nu(D) < \nu(B_1 \backslash A)$ and $\overline{D} \subset B_1 \backslash A$ with $\nu(\overline{D}) = \nu(D)$. Now define $\overline{A} = \overline{D} \cup (A \backslash D)$. It is straightforward to show that

$$\int_{\overline{A}} f(t)d\mu = \int_{\overline{D}} f(t)d\mu - \int_D f(t)d\mu > 0.$$

However, $\nu(\overline{A}) = c$ and so by hypothesis, $\int_{\overline{A}} f(t)d\mu = 0$, a contradiction.

If $\mathrm{supp} f$ is not all of Ω, but has measure greater than c, we can repeat the previous argument with $\mathrm{supp} f$ in place of Ω. If $\nu(\mathrm{supp} f) \leq c$, then there exists a set A with $\nu(A) = c$ and $B_1 \cup B_2 \subset A$. Again, choose $D \subset A \cap B_2$ with $0 < \nu(D) < \nu(\Omega \backslash A)$ and $\overline{D} \subset \Omega \backslash A$ in such a way that $\nu(\overline{D}) = \nu(D)$. If $\overline{A} = \overline{D} \cup (A \backslash D)$, then $\nu(\overline{A}) = \nu(A) = c$ and $\int_{\overline{A}} f(t)d\mu = -\int_D f(t)d\mu > 0$, which is a contradiction. $\square$

We state three integral equalities that will be useful in the proof below.

$$\mu(T^{-1}A) = \int_A g(t)d\mu;$$

$$\int_A fd\mu = \int_{TA} T_1(f)gd\mu;$$

$$\int_{TA} fd\mu = \int_A T_1^{-1}(f[g^{-1}])d\mu;$$

where $f \in L^\Phi$.

5.4.10. THEOREM. *Let $\mu(\Omega_0) = 0$ and suppose that $u \to \frac{\Phi'(u,t)}{u}$ is monotone for each $t \in \Omega$. If U is a surjective isometry on L^Φ, there exist a* *measurable function h, which is different from zero a.e., and a regular set isomorphism T such that*

(105) $Uf(t) = h(t)T_1f(t),$

for all $f \in L^\Phi$ and

(106) $\Phi(\alpha|h(t)|, t) = g(t)T_1\Phi(\alpha, t),$

for almost all $t \in \Omega$ and nonnegative real numbers α.

  Conversely, if a measurable function h, which is different from zero a.e., and a regular set isomorphism T satisfy (106), then an operator U defined by (105) is a surjective isometry of L^Φ.

PROOF. Suppose U is a surjective isometry on L^Φ. Using the characterization of Hermitian operators as given in Theorem 5.4.8 and the fact that $H \rightarrow UHU^{-1}$ sets up an isomorphism on the class of Hermitian operators, we can find, exactly as in the proof of Theorem 5.3.5, a regular set isomorphism T and a measurable function h so that

$$U(\chi_A)(t) = h(t)\chi_{TA}(t)$$

a.e. Then by Theorem 5.3.3 we have (105). Since U is surjective, we must have h different from zero a.e. It remains to show that (106) is satisfied.

  Let $\{B_n\}$ denote an increasing sequence of sets with positive finite measure whose union is Ω and such that $|h|$ is bounded on each B_n. For each n let $D_n = T^{-1}(B_n)$ and choose α_n such that $\int_{D_n} \Phi(\alpha_n, t)d\mu = 1$.

  Suppose $\alpha > \alpha_n$. For $A \subset D_n$ with $\int_A \Phi(\alpha, t)d\mu = 1$, we have

$$\|\alpha\chi_A\| = \|U(\alpha\chi_A)\| = \|\alpha h\chi_{TA}\| = 1.$$

Since h is bounded on $TA \subset B_n$, it must be true that $\alpha h\chi_{TA} \in E^\Phi$ and $I_\Phi(\alpha h\chi_{TA}) = 1$. Therefore, by letting $f(t) = T_1^{-1}(\Phi(\alpha|h(t)|, t)[g(t)]^{-1})$ in the second integral equality preceding the statement of the theorem, we get that

$$\int_A T_1^{-1}([g(t)]^{-1}\Phi(\alpha|h(t)|, t)d\mu = \int_{TA} \Phi(\alpha|h(t)|, t)d\mu = 1.$$

It follows that

$$\int_A \{T_1^{-1}([g(t)]^{-1}\Phi(\alpha|h(t)|, t)) - \Phi(\alpha, t)\}d\mu = 0$$

for all $A \subset D_n$ with $\int_A \Phi(\alpha, t)d\mu = 1$. Now by an application of Lemma 5.4.9 using appropriate identifications, we can conclude that the integrand above is zero a.e in D_n. Hence (106) holds for almost all $t \in D_n$ and $\alpha > \alpha_n$.

  Now suppose $\alpha \leq \alpha_n$, and $\beta = \int_{D_n} \Phi(\alpha, t)d\mu$. Then $0 < \beta \leq 1$ and if $A \subset D_n$ with $\int_A \Phi(\alpha, t)d\mu = \beta/2$, we may choose $\gamma > \alpha_n$ so that

$$\int_{D_n \backslash A} \Phi(\gamma, t)d\mu \geq 1 - \frac{\beta}{2}.$$

Next we select $B \subset D_n \setminus A$ such that $\int_B \Phi(\gamma, t)d\mu = 1 - \beta/2$. Thus,

(107)
$$\int_A \Phi(\alpha, t)d\mu + \int_B \Phi(\gamma, t)d\mu = 1$$

and it follows that

$$\|\alpha\chi_A + \gamma\chi_B\| = 1 \text{ and } \|\alpha h\chi_{TA} + \gamma h\chi_{TB}\| = 1.$$

Once again, the boundedness of h on both TA and TB implies that $\alpha h\chi_{TA} + \gamma h\chi_{TB} \in E^\Phi$ and therefore

$$I_\Phi(\alpha h\chi_{TA} + \gamma h\chi_{TB}) = 1.$$

Using this last equation and the third integral equality preceding the statement of the theorem, we obtain

$$\int_A T_1^{-1}([g(t)]^{-1}\Phi(\alpha|h(t)|, t))d\mu + \int_B T_1^{-1}([g(t)]^{-1}\Phi(\gamma|h(t)|, t))d\mu = 1.$$

Since $\gamma > \alpha_n$, our previous work shows that $\Phi(\gamma|h(t)|, t) = g(t)T_1\Phi(\gamma, t)$ for almost all $t \in D_n$. Hence,

$$\int_A T_1^{-1}([g(t)]^{-1}\Phi(\alpha|h(t)|, t))d\mu + \int_B \Phi(\gamma, t)d\mu = 1.$$

Combining this with (107), we see that

$$\int_A \{T_1^{-1}([g(t)]^{-1}\Phi(\alpha|h(t)|, t)) - \Phi(\alpha, t)\}d\mu = 0$$

for any $A \subset D_n$ with $\int_A \Phi(\alpha, t)d\mu = \alpha/2$. Another application of Lemma 5.4.9 leads us to conclude that (106) holds for all $\alpha \le \alpha_n$ and almost all $t \in D_n$. However, we have already seen that the result holds for $\alpha > \alpha_n$ on D_n, and since $\cup D_n = \Omega$ we have that (106) holds a.e.

For the converse, it is straightforward to show, using both (105) and (106), that

$$I_\Phi(U\chi_A) = I_\Phi(\chi_A).$$

Clearly, the same equation is true for simple functions. For $f \in L^\Phi$, there exists a sequence $\{f_n\}$ of simple functions in E^Φ which converges to f and $|f_n| \le |f|$ a.e. The same holds for $\{\Phi(|f_n(t)|, t)\}$ and $\Phi(|f(t)|, t)$ so by Fatou's Lemma,

$$\int \Phi(|f(t)|, t)d\mu = \liminf \int \Phi(|f_n(t)|, t)d\mu.$$

But $I_\Phi(f_n) = I_\Phi(Uf_n)$ so $I_\Phi(f) = \lim I_\Phi(Uf_n)$. By (105) and properties of T_1, we can apply Fatou's lemma again to get

$$I_\Phi(Uf) = \lim I_\Phi(Uf_n) = I_\Phi(f).$$

Hence, for any $\alpha > 0$, $I_\Phi(\alpha f) = I_\Phi(\alpha Uf)$, from which it follows that $\|Uf\| = \|f\|$. Furthermore, U must be surjective since $h(t) \ne 0$ for almost all $t \in \Omega$. $\square$

5.4.11. REMARKS.

(i) *Theorem 5.4.10 is a generalization of Lumer's Theorem for Orlicz spaces (5.2.12) which includes both necessary and sufficient conditions.*

(ii) *Condition (106) directly generalizes (39) in Lamperti's Theorem 3.2.5.*

(iii) *Theorem 5.4.10 also shows that every isometry on L^Φ is a modular isometry; i.e., $I_\Phi(Uf) = I_\Phi(f)$ for every $f \in L^\Phi$.*

In the introduction, we advertised that we would characterize the isometries on L^Φ for the Orlicz norm and show, in fact, that they are the same as for the Luxemburg norm. In so doing, we intend to be a bit stingy with the details, giving only those which are required to show how to deal with the different norm. We recall that the Orlicz norm is given by

$$\|f\|_0 = \sup \left\{ \int |fg| d\mu : I_{\Phi^*}(g) \leq 1 \right\} = \inf_{k>0} (1 + I_\Phi(kf)).$$

The inequality on the right in the display above is sometimes called the Amemiya formula. In fact the infimum may actually be achieved for some k between $k_1 \leq k_2$ defined by

$$k_1 = \inf \left\{ k > 0 : \int \Phi^*(\Phi'(k|g|)) \geq 1 \right\}$$

$$k_2 = \sup \left\{ k > 0 : \int \Phi^*(\Phi'(k|g|)) \leq 1 \right\}.$$

For each $g \in E^\Phi$, it is possible to associate a function G such that $I_{\Phi^*}(G) = 1$ and such that

$$F_g(f) = \|f\| \int f(t) sgng(t) G(t) d\mu$$

defines a support functional of g. Such functionals would therefore determine a semi-inner product that is compatible with the Orlicz norm. In the case, for example, where $k_1 < k < k_2$, a suitable G is found by taking $G(t) = \Phi'(k|g(t)|, t)$.

We will show that any Orlicz-norm isometry must be almost elementary. To accomplish this, we follow our oft used formula: show that the Hermitian operators are multiplications by bounded real functions. The first step is to prove the following familiar-looking lemma.

5.4.12. LEMMA. *Let H be a Hermitian operator on L^Φ with the Orlicz norm. Let G be a function assigned for $f \in L^\Phi$. For any $A, B \in \Sigma$ with $A \cap B = \emptyset$,*

$$\int_A Hf\chi_B \cdot sgnf \cdot G = \int_B \overline{Hf\chi_A} \cdot sgn\bar{f} \cdot G.$$

PROOF. Let $Pf = f\chi_A$ and $Qf = f\chi_B$. Then P and Q are Hermitian projections on L^Φ satisfying $PQ = 0$. By known results about Hermitian operators, it is the case that for any Hermitian operator H and any $f \in L^\Phi$, the support functional values $F_f(\chi_A Hf\chi_B + \chi_B Hf\chi_A)$ and $iF_f(\chi_A Hf\chi_B - \chi_B Hf\chi_A)$ must be real. Hence,

$$F_f(\chi_A Hf\chi_B) = \overline{F_f(\chi_B Hf\chi_A)}.$$

The result now follows from the definition of F_f. $\qquad\square$

As it has been in the earlier sections of this chapter, the next step is a bit difficult (see Lemma 5.4.7).

5.4.13. LEMMA. *Assume that $\mu(\Omega_0) = 0$. If H is a Hermitian operator on L^Φ with the Orlicz norm, then $\text{supp} H\chi_A \subset A$ almost everywhere for any $A \in \Sigma$.*

PROOF. Let $A \in \Sigma$ with $\mu(\Omega \backslash A) > 0$. Let $0 < \gamma < \beta$ be arbitrary and note that since by our assumptions we have $\int \Phi^*(a\chi_\Omega, t)d\mu < +\infty$ for all $a \geq 0$, we must also have

$$\int \Phi^*(\Phi'(a,t), t)d\mu < +\infty$$

for any $a \geq 0$. Since μ is nonatomic, there exists $A_1 \in \Sigma$ with $A_1 \subset A^c$ such that

$$\mu(A^c \backslash A_1) > 0 \quad \text{and} \quad \int_{A_1} \Phi^*(\Phi'(\beta,t), t)d\mu < 1.$$

Since $\lim_{u \to 0} \Phi'(u,t) = 0$, there exists $\alpha > 0$ such that

$$\int_A \Phi^*(\Phi'_-(\alpha,t), t)d\mu + \int_{A_1} \Phi^*(\Phi'(\beta,t), t)d\mu \leq 1.$$

The α above can be chosen so that we also have

$$\int_A \Phi^*(\Phi'_+(\alpha)) + \int_{A_1} \Phi^*(\Phi'(\beta)) \geq 1.$$

Now it follows that there exists $\lambda_1 \in [0, 1]$ so that

$$\lambda_1 \int_A \Phi^*(\Phi'_-(\alpha) + (1 - \lambda_1) \int_A \Phi^*(\Phi'_+(\alpha)) + \int_{A_1} \Phi^*(\Phi'(\beta)) = 1.$$

Hence by equation (99), we have

$$\int_A \Phi^*(\lambda_1\Phi'_-(\alpha) + (1 - \lambda_1)\Phi_+(\alpha)) + \int_{A_1} \Phi^*(\Phi'(\beta)) = 1.$$

For $f = \alpha\chi_A + \beta\chi_{A_1}$, the function

$$F(t) = (\lambda_1\Phi'_-)\alpha, t) + (1 - \lambda_1)\Phi'_+(\alpha, t))\chi_A(t) + \Phi'(\beta, t)\chi_{A_1}(t)$$

defines a support functional for f.

Since $\gamma < \beta$, we have $\int_{A_1} \Phi^*(\Phi'(\gamma)) \le \int_{A_1} \Phi^*(\Phi'(\beta))$, and in the same way as before, there exist a set (possibly empty) $A_2 \subset (A \cup A_1)^c, \delta > 0$, and $\lambda_2 \in [0,1]$ such that

$$\int_A \Phi^*(\lambda_1 \Phi'_-(\alpha) + (1-\lambda_1)\Phi'_+(\alpha)) + \int_{A_1} \Phi^*(\Phi'(\gamma))$$
$$+ \int_{A_2} \Phi^*(\lambda_2 \Phi'_-(\delta) + (1-\lambda_2)\Phi'_+(\delta)) = 1.$$

If we let $g = \alpha\chi_A + \gamma\chi_{A_1} + \delta\chi_{A_2}$, then

$$G(t) = [\lambda_1\Phi'_-(\alpha,t) + (1-\lambda_1)\Phi'_+(\alpha,t)]\chi_A(t)$$
$$+ [\lambda_2\Phi'_-(\delta,t) + (1-\lambda_2)\Phi'_+(\delta,t)]\chi_{A_2}(t) + \Phi'(\gamma,t)\chi_{A_1}(t)$$

is a support functional for g. We then apply Lemma 5.4.12 to obtain the two equations

$$\int_A H\beta\chi_{A_1} F = \int_{A_1} \overline{H\alpha\chi_A}\Phi'(\beta,t),$$

and

$$\int_A H\gamma\chi_{A_1} G = \int_{A_1} \overline{H\alpha\chi_A}\Phi'(\gamma,t).$$

For $t \in A$, $F(t) = G(t)$ and the above equations may be manipulated in order to obtain

$$(108) \qquad \int_{A_1} \overline{H(\alpha\chi_A)}\left[\frac{\Phi'(\beta,t)}{\beta} - \frac{\Phi'(\gamma,t)}{\gamma}\right] = 0.$$

In fact, we get this same statement for any $B \subset A_1$ and similarly, since $\int_{A^c} \Phi^*(\Phi'(\beta)) < \infty$, we can find disjoint sets $A^1, \ldots, A^k$ with the property that $A^c = \cup_{j=1}^k A^j$ and $\int_{A^j} \Phi^*(\Phi'(\beta)) < 1$ for each j, where also $\mu((A \cup A^j)^c) > 0$. We can repeat the previous arguments with A_1 replaced by A^j and conclude that (108) holds for A^j instead of A_1. Indeed, it follows that (108) holds for any $B \subset A^j$ and hence for any $B \subset A^c$. The integrand in (108) is therefore zero for almost all $t \in A^c$. The conclusion is that $\{t : H\chi_A(t) \ne 0\}$ is contained in Ω_0 which has measure zero, and this implies that $\mathrm{supp}H\chi_A$ is contained in A. $\qquad\square$

We now state formally the facts that the Hermitian operators for the Orlicz norm are multiplications exactly as for the Luxemburg norm, and that the isometries for the Orlicz norm are almost elementary. The proofs are the same as the proofs of Theorem 5.4.8 and Theorem 5.4.10 and will be omitted.

5.4.14. THEOREM. *Suppose that $\mu(\Omega_0) = 0$ where Ω_0 is the set of t for which $\dfrac{\Phi'(u,t)}{u}$ is constant. Suppose further that $u \to \dfrac{\Phi'(u,t)}{u}$ is monotone for every $t \in \Omega$ and that each simple function is in E^Φ.*

(i) *If H is a Hermitian operator on L^Φ with the Orlicz norm, then there is a real valued L^∞-function h such that*

$$Hf = hf$$

for all $f \in L^\Phi$ and $\|H\| = \|h\|_\infty$.

(ii) *If U is a surjective isometry on L^Φ with the Orlicz norm, then there exist a measurable function h, which is different from zero a.e., and a regular set isomorphism T such that*

(109) $$Uf(t) = h(t)T_1 f(t),$$

for all $f \in L^\Phi$.

One of the goals in this section was to show that the isometries are the same for both the Luxemburg and Orlicz norms. Recall from Remark 5.4.11 (iii) that every isometry for the Luxemburg norm is a modular isometry. It follows from the Amemiya formula stated earlier that a modular isometry will be an isometry for the Orlicz norm. In order to verify the converse statement, we will need another lemma.

5.4.15. LEMMA. *Let T_1 be the operator induced by a regular set isomorphism T. If for each $k > 0$ and $v \ge 0$,*

$$\Phi(kv, t) = \frac{\Phi(k, t)}{T_1 \Phi(1, t)} T_1 \Phi(v, t),$$

then

$$T_1 \Phi^*\left(\frac{kuT_1\Phi(1,t)}{\Phi(k,t)}, t\right) = \frac{\Phi^*(u,t)}{\Phi(k,t)} T_1 \Phi(1, t).$$

PROOF. From the definition of Φ^* given by (97) we can write

$$\Phi^*\left(\frac{kuT_1\Phi(1,t)}{\Phi(k,t)}, t\right) = \sup_{v \ge 0}\left[\frac{kuvT_1\Phi(1,t)}{\Phi(k,t)} - \Phi(v,t)\right]$$

$$= \sup_{w \ge 0}\left[uw - \Phi\left(\frac{w\Phi(k,t)}{kT_1\Phi(1,t)}, t\right)\right].$$

Now $\Phi(\lambda, t) = \Phi(\lambda k k^{-1})$ for any $\lambda \ge 0$ and so upon application of the hypotheses and a little manipulation we have

$$T_1 \Phi(\lambda k^{-1}, t) = \frac{\Phi(\lambda, t)T_1\Phi(1,t)}{\Phi(k,t)}.$$

We use this equation and the properties of T_1 to obtain

$$T_1\Phi^*\left(\frac{kuT_1\Phi(1,t)}{\Phi(k,t)},t\right) = T_1\sup_{w\geq 0}\left\{uw - \Phi\left(\frac{\Phi(k,t)wk^{-1}}{T_1\Phi(1,t)},t\right)\right\}$$

$$= \sup_{w\geq 0}\left\{uw - T_1\Phi\left(\frac{\Phi(k,t)wk^{-1}}{T_1\Phi(1,t)},t\right)\right\}$$

$$= \sup_{w\geq 0}\left\{uw - \Phi\left(\frac{\Phi(k,t)w}{T_1\Phi(1,t)}\right)\frac{T_1\Phi(1,t)}{\Phi(k,t)}\right\}$$

$$= \frac{T_1\Phi(1,t)}{\Phi(k,t)}\sup_{w\geq 0}\left\{uw\frac{\Phi(k,t)}{T_1\Phi(1,t)} - \Phi\left(\frac{\Phi(k,t)w}{T_1\Phi(1,t)},t\right)\right\}$$

$$= \frac{T_1\Phi(1,t)}{\Phi(k,t)}\Phi^*(u,t).$$

$$\square$$

5.4.16. THEOREM. *If $\mu(\Omega_0) = 0$, then the groups of isometries of L^Φ in both Luxemburg and Orlicz norms coincide.*

PROOF. We have already observed that every isometry for the Luxemburg norm is an isometry for the Orlicz norm. So assume that U is an isometry for the Orlicz norm. It is known that the adjoint U^* is therefore an isometry for L^{Φ^*} with the Luxemburg norm induced by Φ^*. By Theorem 5.4.10 there is a regular set isomorphism T and a measurable function h so that

$$(U^*g)(t) = h(t)T_1g(t)$$

where

(110) $$\Phi^*(\alpha|h(t)|,t) = w(t)T_1\Phi^*(\alpha,t).$$

The function $w(t)$ satisfies $(\mu \circ T^{-1})(A) = \int_A w(t)d\mu$. Using the third of the integral equalities given just before the statement of Theorem 5.4.10 and the form of U^* above, we have

$$\int (Uf)g = \int (U^*g)f = \int T_1^{-1}([hfT_1g][w]^{-1})$$

$$= \int T_1^{-1}(h)T_1^{-1}(f)T_1^{-1}([w]^{-1})g.$$

From this we conclude that

$$Uf = \frac{T_1^{-1}(h)}{T_1^{-1}(w)}T_1^{-1}(f).$$

By the sufficient conditions given by Theorem 5.4.10 we will be finished if we can show that

(111) $$\Phi\left(\frac{\alpha|T^{-1}(h)|}{T_1^{-1}(w)},t\right) = \frac{1}{T_1^{-1}w(t)}T_1^{-1}\Phi(\alpha,t)$$

Upon letting $\alpha = 1$ in (110), we see that

$$w(t) = \frac{\Phi^*(|h(t)|, t)}{T_1[\Phi^*(1, t)]}.$$

Using these identifications and properties of T_1 and its inverse, one can see that condition (111) is equivalent to

$$T_1\Phi\left(\frac{|h(t)|T_1(\Phi^*(1,t))\alpha}{\Phi^*(|h(t)|, t)}, t\right) = \frac{T_1\Phi^*(1, t)}{\Phi^*(|h(t)|, t)}\Phi(\alpha, t).$$

This equality follows directly from Lemma 5.4.15 applied to Φ^* rather than Φ.   □

We opened this section with a discussion of the Nakano $L^{p(t)}$-spaces. An application of Theorem 5.4.10 immediately yields the following theorem.

5.4.17. THEOREM. *Assume that $p(t)$ is a real measurable function with range in $(1, +\infty)$ and $p(t) \neq 2$ almost everywhere. Then U is a surjective isometry on $L^{p(t)}$ if and only if*

$$Uf = hT_1f$$

for all $f \in L^{p(t)}$, where h is a nonzero measurable function and T_1 is induced by a regular set isomorphism so that

$$T_1p(t) = p(t), \quad g(t) = |h(t)|^{p(t)}$$

where $g(t)$ is the Radon-Nikodym derivative of the measure $(\mu \circ T^{-1})$ with respect to μ.

We close with an example of a nontrivial isometry on a Nakano space. Let T be the set isomorphism on the Lebesgue measurable sets in $[0, 1]$ induced by the point map $\theta(t) = \dfrac{1 - t}{1 + t}$. Let

$$p(t) = \frac{1}{t + \theta(t)} = \frac{1 + t}{1 + t^2},$$

$$h(t) = \left(\frac{2}{(1 + t)^2}\right)^{\frac{1}{p(t)}}.$$

Then $p(t) = p(\theta(t)) = T_1p(t)$ and

$$h(t)^{p(t)} = \frac{2}{(1 + t)^2} = |\theta'(t)|.$$

The conditions of the above theorem are satisfied and

$$Uf(t) = h(t)T_1f(t) = \left(\frac{2}{(1 + t)^2}\right)^{\frac{1}{p(t)}} f\left(\frac{1 - t}{1 + t}\right)$$

is an isometry on $L^{p(t)}[0, 1]$.

5.5. Notes and Remarks

Topics which could be considered as belonging to the study of Banach Function Spaces have been studied ever since the introduction of the Lebesgue L^p-spaces. The primary reference for our discussion has been the book by Bennett and Sharpley [24]. Other good references are [185] and the thesis of Luxemburg [207], where the general notion of Banach function space was introduced and developed.

Lumer's Method for Orlicz Spaces. The properties of Orlicz spaces that we have listed can be found in [24], [185], [200], [227], and [262]. Obviously, there are earlier and original sources for many of the properties, but the reader can chase them down from the references cited. The information about semi-inner products, numerical ranges, and Hermitian operators, as well as part (i) of Lemma 5.2.5 can be found in the 1961 paper of Lumer [203]. The rest of Lemma 5.2.5 and Theorem 5.2.6 are taken from Bonsall and Duncan [42, pp.28,46].

The results from Lemma 5.2.8 through Theorem 5.2.12 are all due to Lumer [204], [205]. Our proofs look slightly different in some places because we have assumed that the norm was absolutely continuous from the outset, and also we have considered φ' to be the left derivative of φ. For the proof of the first part of Lemma 5.2.7 it would be enough to assume $f \in E^{\Phi}$, and the second part can be proved for any $f \in L^{\Phi}$. Lumer's Theorem 5.2.12 gives necessary conditions for U to be an isometry and does not establish conditions for U to be a modular isometry. It is also important in Lumer's method that the isometry be surjective. To get sufficient conditions, one must establish a property for φ like that obtained in equation (106) of Theorem 5.4.10 in Section 4. This was done by Zaidenberg in his paper on groups of isometries on Orlicz spaces [326].

The proof of the theorem for sequence spaces follows that of Arazy [13]. The reader should also consult [300] and [101] for a discussion of this case. See Kaminska's article [163] for a very nice survey of the investigations concerning isometries of Orlicz spaces.

Zaidenberg's Generalization. A rearrangement invariant (r.i. space) or symmetric space is essentially a Banach function space in which functions and their equimeasurable rearrangements have the same norm. Luxemburg [208] may have been the first to give a formal treatment of these spaces. Excellent accounts are given in [24] and [200]. In a series of three papers [326], [327], and [328], Zaidenberg adopted what was essentially Lumer's scheme to produce descriptions of isometries on these generalizations of Orlicz spaces. The 1977 paper [327] contained no proofs; these were given in [328] which was in Russian. More recently, Zaidenberg [329] has published an updated, English translation of that 1980 article, and it is this latest paper that forms the basis for what we have written in the section presently under discussion.

Zaidenberg assumes throughout that the measure space Ω associated with the r.i. spaces is either $[0, 1]$ or a line with Lebesgue measure. Most treatments of r.i. spaces begin with the observation that the study of r.i. spaces over separable (nonatomic) measure spaces reduces to the study of spaces where Ω can be taken to be $[0, 1]$ or $[0, \infty)$ with the usual Lebesgue measure. This idea has been attributed to Luxemburg [24, p.62]. One important consequence of this assumption is that the regular set isomorphism T that occurs in the description of isometries can be given by a point transformation. Such a point transformation is often called an *automorphism* or σ-*automorphism*. The isometries are then always *elementary*, using the language we mentioned in the text and which seems to have been introduced in [161]. In our treatment we have not made this assumption and therefore our characterizations involve the induced operator T_1 and have the form we have called *almost elementary*. This (masochistic?) insistence does contribute to slightly less elegant statements and proofs.

Theorem 5.3.3 and its proof are taken from [327], while Theorems 5.3.4 and 5.3.5 come from [329]. The arguments are difficult and powerful and we have tried to include as much detail as possible. Note that Theorem 5.3.5 gives only necessary conditions for U to be an isometry. Zaidenberg did give a sufficient condition for the Orlicz space case in [326]. As one might expect, the condition involves the Radon-Nidodym derivative of the measure $\nu = \mu \circ \theta^{-1}$ where θ is the σ-automorphism associated with the isometry. A year later, Zaidenberg described the isometries on symmetric spaces and characterized isometry groups in terms of subgroups of the group NS of all automorphisms on the σ-algebra Σ. It is a consequence of the theorem of Banach-Lamperti (Theorem 2.5 of Chapter 3) that when the space $X = L^p$, the isometry group corresponds to all of NS. Zaidenberg showed that the converse is true. It is perhaps worth mentioning here that Abramovich and Zaidenberg have shown that any r.i. space on $[0, 1]$ which is isometric to $L^p[0, 1]$ actually coincides with L^p [2]. Refinements of such questions are treated by Randrianantoanina [261]. Lin [198] has discussed these results in connection with transitivity and maximality of norms on r.i. spaces. This circle of ideas has been summarized nicely by Kaminska [165]. In this paper just cited, Kaminska also discusses isometries of the Calderon-Lozanovskii spaces E_φ where E is a Banach function space and φ a Young's function. By E_φ is meant the class of measurable functions f such that $\varphi(a|f|) \in E$. If $\rho(f) = \|\varphi(|f|)\|_E$ when $\varphi(|f|) \in E$ and $\rho(f) = 0$ otherwise, then

$$\|f\| = \inf\{a > 0 : \rho(f/a) \leq 1\}$$

defines a norm on E_φ with respect to which E_φ is a Banach space. If E is an r.i. space, then E_φ is also an r.i. space [165].

For all of the results in this chapter, it is assumed that the spaces are over the complex scalars. Indeed, Lumer's method requires that the scalars be complex. The real case demands other methods, and these have been supplied by Kalton and Randrianantoanina [161], [162]. As they put it, their methods are "distantly related to the Lumer technique." Their line of reasoning makes use of the notion of *numerically positive operators* and results of Flinn [**269**]. In addition to extending Theorem 5.3.5 to the real case, Kalton and Randrianantoanina showed that if the r.i. space X is not equal to L^p for some $1 \leq p \leq \infty$ up to renorming, then $|h| = 1$ a.e. and T is measure preserving.

Musielak-Orlicz Spaces. The $L^{p(t)}$-spaces are named for Nakano whose work on what he called *modulared spaces* began as early as 1950 [**233**], [**234**], [**235**]. The main reference for general information about such spaces and later developments is the monograph of Musielak [**227**]. The section on comments of this book is an excellent source for the history of work on this subject. The material on isometries of Musielak-Orlicz spaces with respect to the Luxemburg norm is drawn primarily from [**103**]. The assumption that characteristic functions are in the subspace E^Φ was made to make the arguments less cluttered. The assumption is not as restrictive as it might appear, since it is always possible to find a sequence $\{\Omega_n\}$ of measurable sets which form a partition of the measure space $\{\Omega\}$ and such that $\chi_{\Omega_n} \in E^\Phi$ for every n. The other two major assumptions involved the behavior of the function $\dfrac{\Phi'(u,t)}{u}$. The set Ω_0 where this function is constant was assumed to be of measure zero. In fact, the space of functions in L^Φ whose support is contained in Ω_0 is isometric to Hilbert space, so that an isometry U on L^Φ when restricted to this subspace, maps it back into itself. Furthermore, the assumption about monotonicity of $\dfrac{\Phi'(u,t)}{u}$ as a function of u can be removed. The details of these matters can be found in [**144**]. All of the Lemmas and Theorems up through 5.4.10 are taken from [**103**].

The results concerning the Orlicz norm on an M-O space are due to Kaminska [**164**]. The proof of the fact that the Amemiya formula holds can be found in [**185**, p.92]. Musielak [**227**, p.7] refers to this formula as defining the *Amemiya norm*, a reflection of the influence of the Japanese school that grew out of Nakano's pioneering work on modulared spaces. Lemmas and Theorems 5.4.12 through 5.4.16 and their proofs are adapted from similar ones in [**164**].

Once again, we remind the reader that we have assumed that the M-O spaces under discussion are complex Banach spaces. In [**144**], however, the authors are able to prove a version of Theorem 5.4.17 for the real case. The example given to close the section appeared in [**103**], although a misprint in that paper has been corrected.

CHAPTER 6

Banach Algebras

6.1. Introduction

The Banach-Stone theorem tells us that every surjective isometry between $C(Q)$ and $C(K)$ must be of the form $Tf(t) = h(t)f(\varphi(t))$ where φ is a homeomorphism of the compact Hausdorff space K onto the compact Hausdorff space Q, and h is a unimodular function defined on K. The $C(Q)$ spaces are more than just Banach spaces, of course, and are Banach algebras which are, in fact, commutative C^*-algebras. The mapping φ induces an algebra automorphism $\Phi : C(Q) \to C(K)$ which is defined by $\Phi(f) = f \circ \varphi$. The map Φ is more than just an automorphism, it is an isometric-$*$-automorphism in the sense that for every $f \in C(Q)$ it is true that $\Phi(f^*) = (\Phi(f))^*$ (where $f^*(s) \equiv \overline{f}(s)$) and $\|\Phi(f)\| = \|f\|$. In addition to the automorphism, there is multiplication by an element h of $C(K)$ which satisfies the condition $hh^* = h^*h = 1$. Hence, every linear isometry in this setting is the action of a $*$-*automorphism* followed by multiplication by a *unitary element*.

Given the Banach-Stone result for the commutative C^*-algebra $C(Q)$, it is natural to attempt an extension to the noncommutative case. This would mean the abandonment of the function space setting that has occupied us in previous chapters. The first generalization to the noncommutative case was obtained by R. Kadison in 1951 who stated and proved the following theorem.

6.1.1. THEOREM. *(Kadison) A linear isomorphism T of one C^*-algebra $\mathcal{A}$ with identity onto another $\mathcal{B}$ which is isometric, is a C^*-isomorphism followed by a left multiplication by a fixed unitary operator, viz. $T(I)$.*

In Kadison's terminology, a C^*-isomorphism between C^*-algebras $\mathcal{A}$ and $\mathcal{B}$ is an isomorphism of $\mathcal{A}$ onto $\mathcal{B}$ which preserves the self adjoint elements and their power structure (and hence the Jordan structure of $\mathcal{A}$). In the commutative case we used the fact that the adjoint mapping T^* must take extreme points of the dual ball to extreme points of the dual ball. Kadison made use of the fact that T must preserve extreme points of the unit ball of $\mathcal{A}$ itself. Our proof of the theorem, to be given in Section 2, follows a proof given by Paterson, but it also uses the fact that extreme points must go to extreme points under a surjective isometry. Kadison's theorem gave rise to a burst of activity involving J^*-algebras and JB^*-triples which we will discuss briefly in the Remarks at the end of the chapter. Section 3 is devoted to still another proof of Kadison's Theorem, an effort of W. Werner using a differentiability condition which stirs echoes of Banach's original proof as given in Chapter 1. In Section 4, we will follow the pattern of earlier chapters by considering the nature of an isometry between C^*-algebras which is not surjective. Our treatment will be given in the setting of noncommutative functional analysis and will consider complete isometries.

In the remainder of the chapter, we return to function spaces. In Section 5 we examine Cambern's treatment of the isometries on the algebras $C^{(1)}([0,1])$ and $AC([0,1])$. The arguments here will once again make use of the knowledge of the extreme points in the dual balls. Finally, we discuss some fairly recent work of Font that includes a characterization of the isometries on Douglas algebras; that is, closed subalgebras of L^∞ which contain H^∞.

6.2. Kadison's Theorem

The extension of the Banach-Stone Theorem to the noncommutative case was extremely significant, and its appearance was followed by a flurry of activity that is still going on today. Kadison showed that the set of extreme points of the unit ball of a C^*-algebra $\mathcal{A}$ with identity is exactly the set of partially isometric elements of $\mathcal{A}$ of the form u, where $u^*u = p$, $uu^* = q$ and $(1-q)\mathcal{A}(1-p) = \{0\}$. The only normal extreme points are the unitary elements in $\mathcal{A}$ and these are the only extreme points with inverses. He then used this characterization and the fact that isometries must preserve extreme points to prove his theorem. As with any important theorem, new proofs are discovered by later investigators and we are drawn to the interesting proof of Paterson. We will not use the explicit description of extreme points given above, but extreme points will show up in the arguments.

By a C^*-algebra, we mean a Banach-$*$-algebra for which $\|x^*x\| = \|x\|^2$, sometimes called an *abstract C^*-algebra* or a B^*-*algebra*. Of course, by the Gelfand-Neumark theory, a commutative C^*-algebra with unity is represented by $C(Q)$ for a suitable compact Hausdorff space Q, and a noncommutative C^*-algebra with unity is represented by a closed, self-adjoint subalgebra of the bounded linear operators on a Hilbert space. This latter representation is sometimes given as the definition of a C^*-algebra. These well-known representations often provide a good guide for arguments in the abstract setting.

In Section 5.2, we considered the notion of the numerical range of an operator on a Banach space, and defined what was meant by a Hermitian operator. These concepts are very natural for Banach algebras. Let $\mathcal{A}$ be a unital Banach algebra. Although there is potential for confusing the $*$ denoting the involution and the same symbol for the dual of a Banach space, we trust the reader will be able to sort that out. Given an element $a \in \mathcal{A}$, we define the set $D(a; \mathcal{A})$ by

$$D(a; \mathcal{A}) = \{f \in S(\mathcal{A}^*) : f(a) = \|a\|\}.$$

Frequently we will suppress the $\mathcal{A}$ in the notation above and just write $D(a)$. Next let

$$V(a; \mathcal{A}) = \{f(a) : f \in D(a; \mathcal{A})\},$$

and

$$v(a) = \sup\{|\lambda| : \lambda \in V(a; \mathcal{A})\}.$$

The set $V(a; \mathcal{A})$ is called the *numerical range* of a, and $v(a)$ is called the numerical radius. An application of the Hahn-Banach theorem shows that if $\mathcal{B}$ is any subalgebra of $\mathcal{A}$ containing 1, then $V(a; \mathcal{A}) = V(a; \mathcal{B})$ and so we will often just write $V(a)$. An important fact about the numerical radius when the field is complex is given by

(112) $$\|a\| \geq v(a) \geq \frac{1}{e}\|a\|.$$

An element $a \in \mathcal{A}$ is said to be *Hermitian* if and only if $V(a) \subset \mathbb{R}$, and we let $H(\mathcal{A})$ denote the collection of Hermitian elements of $\mathcal{A}$. The next theorem lists some facts about Hermitian elements, and we state it without proof. Some parts are equivalent to those of Theorem 5.2.6. In the statement, $r(x)$ denotes the spectral radius of x.

6.2.1. THEOREM. *Let $\mathcal{A}$ be a unital Banach algebra. Then the following are equivalent:*

(i) $x \in H(\mathcal{A})$,
(ii) $\lim_{t \to 0} t^{-1}(\|1 - itx\| - 1) = 0$,
(iii) $\|e^{itx}\| = 1$,
(iv) $r(x + i\alpha) = \|x + i\alpha\|$ *for every $\alpha \in \mathbb{R}$.*
    Furthermore,
(v) *If $x, y \in H(\mathcal{A})$, then $i(xy - yx) \in H(\mathcal{A})$.*
    If $\mathcal{A}$ is a C^*-algebra then
(vi) $H(\mathcal{A}) = \{a \in \mathcal{A} : a = a^*\}$.

Suppose that f is a nonnegative continuous function on a compact Hausdorff space Q with $\|f\|_\infty \leq 1$, which is not an idempotent. Then there exists some $s_0 \in Q$ such that $0 < f(s_0) < 1$. There is a compact neighborhood W of s_0 such that $\sup f(s) < 1$ on W and by Urysohn's lemma, there exists a $g \in C(Q)$ with range in $[0, 1]$ such that $g(s_0) = 1$ and $g(s) = 0$ for s not in W. We may choose $\epsilon > 0$ and sufficiently small that $f(1 \pm \epsilon g)$ is nonnegative and is in the unit ball of $C(Q)$. This argument can be adapted to prove the following lemma.

6.2.2. LEMMA. *Let $\mathcal{A}$ be a commutative C^*-algebra. If x is a nonnegative element of the unit ball of $\mathcal{A}$ which is not an idempotent, then there exists a nonnegative element a in the unit ball of $\mathcal{A}$ for which*

$$xa \neq 0, \quad \text{and} \quad \|x(1 \pm a)\| \leq 1.$$

Note that if A is a unital normed algebra, and if $f(a) = 0$ for all $f \in D(1; A)$, then $a = 0$ by (112). (This property is sometimes described by saying that 1 is a *vertex* of the unit ball of A.) Suppose we write

$$1 = \frac{1}{2}(1 - a) + \frac{1}{2}(1 + a)$$

where $a \in A$ is such that $\|1 \pm a\| \leq 1$. Because the number 1 is an extreme point of the unit disk in the plane, we must have $f(a) = 0$ for all $f \in D(1)$. By the remark just above, we conclude that $a = 0$ and 1 is an extreme point of the unit ball of A. This is a very special case of the next lemma.

Before stating this lemma, let us introduce some common terminology for elements of a C^*-algebra A. An element $x \in A$ is *self adjoint* is $x = x^*$; *normal* if $x^*x = xx^*$; *unitary* if $x^*x = xx^* = 1$; a *projection* if $x^2 = x$ and $x^* = x$.

6.2.3. LEMMA. *A unitary element of a unital C^*-algebra A is an extreme point of the unit ball.*

PROOF. Suppose $x \in A$ is unitary and $x = \frac{1}{2}(y + z)$ for $y, z \in S(A)$. Substituting this expression in the equation $x^*x = 1$ yields the convex combination

$$1 = \frac{1}{2}\left[\frac{y^*y + y^*z}{2} + \frac{z^*y + z^*z}{2}\right]$$

and a similar expression is obtained by using the equation $xx^* = 1$. From the fact that 1 is an extreme point, we are able to conclude that y and z are themselves unitary and that $y^*z = 1$. Hence,

$$y = y(y^*z) = (yy^*)z = z,$$

which completes the proof.   □

6.2.4. LEMMA. *Let A and B be unital C^*-algebras and T a surjective isometry from A onto B. If u is a unitary element of A, then Tu is a unitary element of B.*

PROOF. Suppose that $u \in A$ and $uu^* = u^*u = 1$. We first prove that Tu is not a zero divisor. Let $x \in B$ be an element for which $(Tu)x = 0$. Since T is surjective, there exists $y \in B$ such that $x = (Ty)^*$. Thus, $(Tu)(Ty)^* = (Ty)(Tu)^* = 0$. For any $\alpha \in \mathbb{C}$,

$$\|u + \alpha y\|^2 = \|Tu + \alpha Ty\|^2 = \|(Tu + \alpha Ty)(Tu + \alpha Ty)^*\|,$$

so that

$$\|u + \alpha y\|^2 = \|(Tu)(Tu)^* + |\alpha|^2(Ty)(Ty)^*\| \leq \|Tu\|^2 + |\alpha|^2\|Ty\|^2.$$

This gives $\|u + \alpha y\| \leq (1 + |\alpha|^2\|Tu\|^2)^{1/2}$. Since u is unitary, it follows that

$$\|1 + \alpha u^*y\| = \|u + \alpha y\| \leq (1 + |\alpha|^2\|Ty\|^2)^{1/2}.$$

Now by letting $\alpha \to 0$ for $\alpha \in \mathbb{R}$, we observe that

$$\|1 + \alpha u^* y\| \leq 1 + o(\alpha) \quad \text{and} \quad \|1 + i\alpha u^* y\| \leq 1 + o(\alpha).$$

Therefore, by Theorem 6.2.1, we conclude that $u^* y \in H(\mathcal{A}) \cap iH(\mathcal{A}) = \{0\}$. Hence, $y = 0$, so that $x = (Ty)^* = 0$. This shows that Tu is not a left divisor of zero and a similar argument would show it is not a right divisor of zero either.

Now we want to prove that Tu is unitary. By Lemma 6.2.3, u is an extreme point, and since T is a surjective isometry, $x = Tu$ must itself be an extreme point of the unit ball of $\mathcal{B}$. Now $x^* x$ must be idempotent, for if that is not the case, then $|x| = \sqrt{x^* x}$ is also not an idempotent. If we consider $|x|$ as a nonnegative element of the commutative C^*-algebra $\mathcal{B}_x$ generated by $|x|$ and 1, then by Lemma 6.2.2, there is a nonnegative element a in the unit ball of $\mathcal{B}_x$ such that $|x|a \neq 0$, and $\||x|(1 \pm a)\| \leq 1$. Since the algebra we are working with here is abelian, and the elements $|x|$ and a are self adjoint, we see that

$$\|[(1 \pm a)|x|][(1 \pm a)|x|]^*\| = \|(1 \pm a)x^* x(1 \pm a)\|.$$

Therefore,

$$\|(1 \pm a)|x|\| = \|(1 \pm a)(x^* x)(1 \pm a)\|^{(1/2)} = [\|x(1 \pm a)\|^2]^{(1/2)}.$$

Thus $\|x(1 \pm a)\| \leq 1$ and $\|xa\| = \|a|x|\| \neq 0$, and we would conclude that $x = (1/2)[x(1 + a) + x(1 - a)]$ is not extreme. Since $x^* x$ is idempotent, and neither x nor x^* is a zero divisor, we must have $x^* x = 1$. Since $x^* = (Tu)^*$ is also an extreme point, we have $xx^* = 1$ also, and $x = Tu$ is unitary. $\square$

We are now ready to prove Kadison's theorem which we state again as follows.

6.2.5. THEOREM. *Let $\mathcal{A}$ and $\mathcal{B}$ be unital C^*-algebras. Then T is a linear isometry from $\mathcal{A}$ onto $\mathcal{B}$, if and only if there exists a unitary element v of $\mathcal{B}$ and a C^*-isomorphism τ such that $T = v\tau$.*

PROOF. We prove the sufficiency first. To that end, let $T = v\tau$ as described in the statement. For any $a \in \mathcal{A}$ we have

$$\|Ta\|^2 = \|(Ta)^*(Ta)\| = \|(v\tau(a))^*(v\tau(a))\|$$
$$= \|\tau(a)^* \tau(a)\| = \|\tau(a^*)\tau(a)\| = r(\tau(a^* a))$$
$$= r(a^* a) = \|a^* a\| = \|a\|^2.$$

For the necessity, let T be an isometry as given in the theorem. Set $v = T1$. Then v is unitary by Lemma 6.2.4 and so if we let $\tau = v^* T$, we have T in the proper form. It remains to show that τ is a C^*-isomorphism.

From the definition τ is linear from $\mathcal{A}$ to $\mathcal{B}$, and since v is unitary, $\|\tau(a)\| = \|v^* T(a)\| = \|T(a)\| = \|a\|$. Hence, τ is an isometry and also $\tau(1) = 1$. If $h \in H(\mathcal{A})$ then,

$$\|1 + i\alpha\tau(h)\| = \|\tau(1) + i\alpha\tau(h)\| = \|1 + i\alpha h\| \leq 1 + o(\alpha)$$

as $\alpha \to 0$. Therefore $\tau(H(\mathcal{A})) \subset H(\mathcal{A})$ and consequently $\tau(a^*) = (\tau(a))^*$. For $h \in H(\mathcal{A})$ and $\alpha \in \mathbb{R}$, the continuity of $a \to a^*$ implies that $e^{i\alpha h}$ is a unitary element in $\mathcal{A}$ and hence $\tau(e^{i\alpha h})$ is a unitary element of $\mathcal{B}$. Since τ is continuous and $\tau(1) = 1$, it follows that $\tau(e^{i\alpha h})\tau(e^{-i\alpha h}) = 1$. Then we must have $[1 + i\alpha\tau(h) - \alpha^2\tau(h^2)/2][1 - i\alpha\tau(h) - \alpha^2\tau(h^2)/2] = 1 + O(\alpha^3)$. This in turn implies that $1 + \alpha^2[(\tau(h)^2 - \tau(h^2)] = 1 + O(\alpha^3)$ as $\alpha \to 0$. It follows immediately that $\tau(h^2) = (\tau(h))^2$. To see that this holds for general elements $x \in \mathcal{A}$, simply write $x = h + ik$ where h, k belong to $H(\mathcal{A})$. By the preceding argument , $\tau((h + k)^2) = (\tau(h + k))^2$ from which it follows that $\tau(hk + kh) = \tau(h)\tau(k) + \tau(k)\tau(h)$. Finally we see that $\tau(x^2) = \tau(h^2 - k^2 + i(hk + kh)) = (\tau(x))^2$, and thus we have shown that τ is a C^*-isomorphism. This completes the proof.   □

We want to note here that the C^*-isomorphism τ is not necessarily a true algebra isomorphism for it may not preserve products. It does preserve the so called Jordan product, $x \circ y = \dfrac{xy + yx}{2}$. For $x, y, z \in \mathcal{A}$, if we define

$$\{x, y, z\} = \frac{1}{2}(xy^*z + zy^*x),$$

then

$$\{x, y, z\} = (x \circ y^*) \circ z + (z \circ y^*) \circ x - (x \circ z) \circ y^*$$

and τ preserves this *Jordan triple product*. In fact, the isometry T itself preserves this triple product, and suggests generalizations of Kadison's theorem to spaces called JB^* triple systems.   These are spaces on which there is a suitably defined triple product $\{\cdot, \cdot, \cdot\}$, a $*$-operation, and a norm which satisfy certain properties. A JB^* homomorphism (or isomorphism) is a complex linear and bounded mapping which respects the Jordan product and the $*$-operation. Two such triple systems are then isometrically isomorphic as Banach spaces if and only if they are isomorphic as JB^* triple systems. See the notes at the end of the chapter for more on these ideas.

In our work above, we assumed that the C^*-algebra was unital. A commutative C^*-algebra without identity can be representated as a $C_0(Q)$ space, where Q is locally compact, and the isometries in this case have been characterized as we saw in Chapter 2. In the noncommutative case, the relevant theorem is the following.

6.2.6. THEOREM. *(Paterson and Sinclair) Let $\mathcal{A}$ and $\mathcal{B}$ be C^*-algebras and T a linear isometry from $\mathcal{A}$ onto $\mathcal{B}$. Then there is a JB^* (that is, a C^*-isomorphism) τ from $\mathcal{A}$ onto $\mathcal{B}$ and a unitary element U in the algebra $\mathfrak{L}(\mathcal{B})$ such that $T = U\tau$.*

By $\mathfrak{L}(\mathcal{B})$ is meant the set of operators R on $\mathcal{B}$ such that there exists an operator S on $\mathcal{B}$ with $aR(b) = S(a)b$. The pair (R, S) is called a *double centralizer* on $\mathcal{B}$. We are not going to discuss the proof of this theorem here.

6.3. Subdifferentiability and Kadison's Theorem

In Section 1.2 we gave Banach's proof for his version of the Banach-Stone Theorem. In that proof Banach used the Gâteaux differentiability of the norm to identify peak points which then corresponded to other peak points under the isometry. Norm conditions are useful since they are likely to be preserved by an isometry. In this section, we are going to use a differentiability condition on the norm of a C^*-algebra to help find the C^*-isomorphism that arises in the description of the isometry in Kadison's theorem. We begin with a result for von Neumann algebras that will provide the means for showing we have the necessary isomorphism. A *von Neumann* algebra is usually defined as a C^*-subalgebra of the bounded operators on a Hilbert space which is closed in the weak operator topology. More abstractly it has been defined as a C^*-algebra with a predual. Another defining property sometimes used is that the algebra is equal to its bicommutant.

6.3.1. LEMMA. *(W. Werner) A bounded linear map Ψ from one von Neumann algebra $\mathcal{A}$ onto a von Neumann algebra $\mathcal{B}$ is a JB^*-homomorphism if and only if Ψ maps the projections in $\mathcal{A}$ to projections in $\mathcal{B}$. The map is injective if and only if $\Psi(p) \neq 0$ for all nonzero projections p.*

PROOF. We begin by assuming that Ψ sends projections to projections. Suppose that p, q are projections with $pq = qp = 0$. Then $p+q$ is a projection, and by hypothesis, $[\Psi(p) + \Psi(q)]^2 = \Psi(p) + \Psi(q)$. It follows that

$$\Psi(p)\Psi(q) = -\Psi(q)\Psi(p)$$

from which we obtain (upon multiplying appropriately on both sides by first $\Psi(p)$ on the left and then on the right by $-\Psi(p)$)

$$\Psi(p)\Psi(q) = -\Psi(p)\psi(q)\Psi(p) = \Psi(q)\Psi(p).$$

Hence, we must have $\Psi(p)\Psi(q) = 0$. Now by the spectral theorem, any self-adjoint element of $\mathcal{A}$ is the norm-limit of a sequence of elements of the form $\sum_{j=1}^{n} \alpha_j p_j$ for some real scalars α_j and mutually orthogonal projections p_j, i.e., $p_j p_k = 0$ if $j \neq k$. It follows from above that $\Psi(a^2) = \Psi(a)^2$ and $\Psi(a)^* = \Psi(a)$ for any self-adjoint element a. The extension to arbitrary elements x so that $\Psi(x^2) = \Psi(x)^2$ and $\Psi(x^*) = \Psi(x)^*$ is carried out in the usual way.

If $\Psi(p) \neq 0$ for all projections $p \neq 0$, then Ψ is injective on all elements of the form $\sum \alpha_j p_j$ as above, and since $\| \sum_{j=1}^{n} \alpha_j p_j \| = \max|\alpha_j|$, the restriction of Ψ to the self-adjoint elements must be injective. The extension to arbitrary elements follows because Ψ preserves the $*$-operation.

The converse follows easily since the properties of a JB^*-homomorphism clearly implies it must take projections to projections.   □

Let us now give the definition of the differentiability condition.

6.3.2. DEFINITION. *A real-valued function ψ on a Banach space X is said to be* strongly subdifferentiable *at a point $x \in X$ whenever the limit*

$$\lim_{t \to 0^+} \frac{\psi(x + tu) - \psi(x)}{t}$$

exists uniformly for $u \in S(X)$, the unit sphere of X.

An element v of a C^*-algebra is called a *partial isometry* if v^*v is a projection. In this case, vv^* is also a projection. Given any element a, there is a partial isometry v such that $a = v|a|$, where $v^*va = a$ and $|a| = \sqrt{a^*a}$. This is called the *polar decomposition* of a.

Our purpose in this section is to give an idea of how the notion of sub-differentiability of the norm at certain elements can be used to describe an isometry acting on the spaces. We will need some technical results which we state below. In some cases we will omit the proofs or provide bare sketches. The remarks at the end of the chapter will guide the reader in finding the details.

Recall that $D(x)$ is the set of functionals φ in $S(X^*)$ for which $\varphi(x) = \|x\|$. Suppose that the norm is strongly subdifferentiable at x and for each $y \in X$ let

$$f(y) = \lim_{t \to 0^+} \frac{\|x + ty\| - \|x\|}{t}.$$

Then f is a real sublinear functional and

$$(113) \qquad\qquad f(y) \leq \frac{\|x + ty\| - \|x\|}{t}$$

for all $t > 0$. It is, in fact, the case that

$$(114) \qquad\qquad f(y) = \max\{\Re x^*(y) : x^* \in D(x)\}$$

for all $y \in X$. It is straightforward to show that $x^* \in D(x)$ if and only if $\Re x^*(y - x) \leq \|y\| - \|x\|$ for all $y \in X$. To see that (114) holds, note first that for $x^* \in D(x)$, we have

$$\Re x^*(y) = \frac{1}{t}\Re x^*[(x + ty) - x] \leq \frac{\|x + ty\| - \|x\|}{t}$$

for all $t > 0$, so that $\Re x^*(y) \leq f(y)$. On the other hand, for a given y we can let $z^*(y) = f(y)$ and use the Hahn-Banach theorem to extend to a real linear functional w^* on X for which $w^*(z) \leq f(z) \leq \|z\|$ for all $z \in X$. Let x^* be the bounded linear functional whose real part is w^*. Then from (113) we obtain

$$\Re x^*(z - x) \leq f(z - x) \leq t^{-1}[\|x + t(z - x)\| - \|x\|]$$

and so upon letting $t = 1$, we get $\Re x^*(z - x) \leq \|z\| - \|x\|$ for all z and we conclude that $x^* \in D(x)$. This proves equation (114).

6.3.3. LEMMA. *(Gregory) The norm of X is strongly subdifferentiable at $x \in S(X)$ if and only if the distance $d(\varphi_n, D(x))$ tends to zero for any sequence $\{\varphi_n\}$ in the dual unit ball such that $\Re \varphi_n(x) \to 1$.*

PROOF. First we prove the "only if" part. Suppose the norm is strongly subdifferentiable at x, and φ_n is a sequence in $B(X^*)$ such that $\Re\varphi_n(x) \to 1$. Let $\epsilon > 0$ be given and choose t as in Definition 6.3.2 so that

$$\left| \frac{\|x + ty\| - \|x\|}{t} - f(y) \right| < \frac{\epsilon}{2}$$

for all $y \in S(X)$. Let $0 < \eta < t\epsilon/2$, and suppose $y^* \in X^*$ satisfies

$$\Re y^*(z - x) \leq \|z\| - \|x\| + \eta.$$

Given $y \in S(X)$, choose $z = x + ty$ in the above inequality to obtain

$$\|x + ty\| \geq \|x\| + t\Re y^*(y) - \eta.$$

From (114) we can find $x^* \in D(x)$ such that $x^*(y) = f(y)$. It now follows that

$$\begin{aligned}
\Re(y^* - x^*)(y) &\leq \frac{\|s + ty\| - \|x\|}{t} + \frac{\eta}{t} - x^*(y) \\
&\leq \frac{\|x + ty\| - \|x\|}{t} - f(y) + \frac{\epsilon}{2} \\
&< \epsilon.
\end{aligned}$$

Hence,

$$\inf\{\Re(y^* - x^*)(y) : x^* \in D(x)\} < \epsilon$$

for each $y \in S(X)$. The set $\epsilon^{-1}(y^* - D(x))$ is weak*-closed and convex, and by what we have established above, the sets $\epsilon^{-1}(y^* - D(x))$ and $B(X^*)$ must have nonempty intersection. Otherwise, we get a contradiction of the separation theorem for compact convex sets. Hence, $y^* \in [D(x) + \epsilon B(X^*)]$.

Since $\Re\varphi_n(x) \to 1$, given η as in the argument above, we can choose n so that $|\Re\varphi_n(x) - 1| < \eta$. From this we have

$$\Re\varphi_n(y - x) \leq \|y\| - \|x\| + \eta, \quad \text{for all } y \in X.$$

By the argument in the previous paragraph, $\varphi_n \in D(x) + \epsilon B(X^*)$, which establishes the desired result.

For the "if" part, we first note that given any $y \in X$, we have that

$$\lim_{t \to 0^+} \frac{\|x + ty\| - \|x\|}{t} = f(y) \text{ exists.}$$

Furthermore, it follows that for all $x^* \in D(x)$ and $(x + ty)^* \in D(x + ty)$ we have

$$\begin{aligned}
0 &\leq t^{-1}[\|x + ty\| - \|x\|] - f(y) \\
&\leq [\Re(x + ty)^* - \Re x^*](y) \\
&\leq \|(x + ty)^* - x^*\|.
\end{aligned}$$

The goal is to show that the right hand side above goes to zero as t goes to zero through positive values.

If we denote by $D_\eta(z)$ the set of $y^* \in X^*$ for which

$$\Re y^*(w - z) \leq \|w\| - \|z\| + \eta,$$

then our argument in the first part actually shows that for any $\epsilon > 0$, there is an $\eta > 0$ such that $D_\eta(x) \subset D(x) + \epsilon B(X^*)$. It can be shown, in fact, that there exist positive numbers γ, ϵ such that $D_\gamma(y) \subset D_\eta(x)$ whenever $\|y - x\| < \epsilon$. Given $\eta > 0$ choose $\gamma < \eta/2$. Then for $y^* \in D_\gamma(y)$ for a given y, and any $z \in X$,

$$
\begin{aligned}
\|z\| - \|x\| &\geq \Re y^*(z - y) - \gamma + \|y\| - \|x\| \\
&\geq \Re y^*(z - x) - \gamma + \Re y^*(x - y) + \|y\| - \|x\| \\
&\geq \Re y^*(z - x) - 2\gamma + (\|y\| - \|2y - x\|) + \|u\| - \|x\|.
\end{aligned}
$$

If we choose $\epsilon = (\eta - 2\gamma)/2$, it follows that

$$\|z\| - \|x\| \geq \Re y^*(z - x) - \eta,$$

so that $y^* \in D_\eta(x)$.

Suppose now that $\{t_n\}$ is a sequence of positive numbers going to zero. For n sufficiently large, we have $t_n < \epsilon$, so that for any $y \in S(X)$, and any $\gamma < \eta/2$, $D_\gamma(x + t_n y) \subset D_\eta(x)$. Let $y \in S(X)$ be given. For each n let $\varphi_n \in D(x + t_n y)$. Since $\varphi_n \in D_\gamma(x + t_n y)$ for any γ, we get $Re\varphi_n(z - x) \leq \|z\| - \|x\| + \eta$ for all z whenever $t_n < \epsilon$. (Note the choice of n is independent of y.) Recalling that $\|x\| = 1$, and η is arbitrary, it is easily shown that $\Re\varphi_n(x) \to 1$. The condition now gives that $d(\varphi_n, D(x)) \to 0$ and this leads to the desired goal mentioned above. $\qquad\square$

6.3.4. LEMMA. *(Taylor and Werner) Let $\mathcal{A}$ be a C^*-algebra and $\{\varphi_n\}$ a sequence in $B(\mathcal{A}^*)$. Suppose that p, q are projections in $\mathcal{A}^{**}$ such that $\|p\varphi_n q\| \to 1$. Then $\|p\varphi_n q - \varphi_n\| \to 0$.*

Here it is necessary to interpret the meaning of expressions of the form $p\varphi$ and φq. Since $q \in \mathcal{A}^{**}$ and $\varphi \in \mathcal{A}^*$, we understand φq to be the element of $\mathcal{A}^*$ defined at $b \in \mathcal{A}$ by $(\varphi q)(b) = (q\hat{b})(\varphi)$. Similarly, $p\varphi$ is given by $(p\varphi)(b) = (\hat{b}p)(\varphi)$.

6.3.5. LEMMA. *(Contreras, Paya, and Werner) Suppose the norm of $\mathcal{A}$ is strongly subdifferentiable at $a \in S(\mathcal{A})$, and v is a partial isometry in $\mathcal{A}^{**}$ with $va \in \mathcal{A}$ and $v^*va = a$. Then the norm is strongly subdifferentiable at va.*

PROOF. We will make use of Lemma 6.3.3. To that end, let $\{\varphi_n\}$ be a sequence in $S(\mathcal{A}^*)$ with $\varphi_n(va) \to 1$. If we let $\varphi_n' = \varphi_n v$, then the hypothesis and the Lemma provides a sequence $\{\psi_n'\}$ in $D(a)$ so that $\|\varphi_n' - \psi_n'\| \to 0$. Let $\psi_n = \psi_n' v^*$ and note that each $\psi_n \in D(va)$. Now

$$\|\varphi_n vv^* - \psi_n\| = \|\varphi_n' v^* - \psi_n' v^*\| \leq \|\varphi_n' - \psi_n'\| \to 0,$$

and also $\|\varphi_n vv^* - \varphi_n\| \to o$. From this we get that $\|\varphi_n - \psi_n\| \to 0$, and by the Lemma again, the norm is subdifferentiable at va. $\qquad\square$

6.3.6. THEOREM. *(Contreras, Paya, and Werner) Let $\mathcal{A}$ be a C^*-algebra and $\alpha \in S(\mathcal{A})$. The following are equivalent.*

(i) *The norm of $\mathcal{A}$ is strongly subdifferentiable at a.*

(ii) 1 *is an isolated point in the spectrum* $sp(|a|)$ *of* $|a|$.
(iii) *There exists a partial isometry* $v \in \mathcal{A}$ *such that*

$$a \in F_{v,0} = \{x \in S(\mathcal{A}) : xv^* = vv^*, \|x - v\| < 1\}.$$

PROOF. (i) $\implies$ (ii). Let $a \in \mathcal{A}$. By the polar decomposition and Lemma 6.3.5, we conclude that the norm is strongly subdifferentiable at $|a|$. If $\mathcal{B}$ denotes the commutative C^*-algebra generated by $|a|$, then $\mathcal{B}$ can be represented by a space $C_0(Q)$, where Q is the locally compact space $sp(|a|)\backslash\{0\}$. It can be shown that $D(|a|, \mathcal{B}) = \{\psi_1\}$, where, as usual, ψ_t is the evaluation functional at t. If we assume that 1 is not isolated in the spectrum, we can show the norm is not subdiffferentiable at $|a|$ by using Lemma 6.3.3. This contradiction establishes the implication.

(ii) $\implies$ (iii). One first uses the continuous functional calculus to find a projection $p \in \mathcal{A}$ such that $|a| \in F_{p,0}$. If $a = u|a|$ is the polar decomposition of a, then $v = up$ can be shown to satisfy the condition in (iii).

(iii) $\implies$ (i). For $p = v^*v$, it is possible to show that $|a| \in F_{p,0}$ and then that the norm is strongly subdifferentiable at $|a|$. The key is to show that $D(|a|) = D(p)$, which is done by considering the commutative C^*-algebra generated by $|a|$ and p. This algebra can be represented by a $C_0(Q)$ and it is not too difficult to show that $D(p, C_0(Q)) = D(|a|, C_0(Q))$. From Lemma 6.3.5, we get subdifferentiability at a. We omit the details. $\qquad\square$

We need one more technical lemma.

6.3.7. LEMMA. *(Werner) Let* v *be a partial isometry in* $\mathcal{A}$ *and* $a \in S(\mathcal{A})$. *Then*

(i) $a \in F_{v,0}$ *if and only if* $D(a) = D(v)$.
(ii) *If* $x \in F_{v,0}$ *satisfies* $\|x - a\| < 1$ *for all* $a \in F_{v,0}$, *then* $x = v$.

PROOF. (i) Given $a \in F_{v,0}$, from Theorem 6.3.6, 1 is an isolated point in the spectrum of $|a|$ and from this one can show that for any projection p satisfying $|a|p = p$ and $\||a| - p\| < 1$, then $p = v^*v$. Since $v = vv^*v = av^*v = ap$, we see that there can be only one partial isometry with $av^* = vv^*$ and $\|a - v\| < 1$. If $a = u|a|$ is the polar decomposition of a, then $a \in F_{up,0}$ as mentioned in the proof of (ii) implies (iii) of 6.3.6. Hence by the uniqueness, we must have $v = up$. We have seen in the proof of (iii) $\implies$ (i) in the previous theorem that $D(|a|) = D(p)$. If $\varphi(v) = 1$, then $\varphi u(p) = \varphi(up) = \varphi(v) = 1$, and so $\varphi(a) = \varphi(u|a|) = \varphi u(|a|) = 1$. Similarly, $\varphi(v) = 1$ whenever $\varphi(a) = 1$ and we see that $D(a) = D(v)$.

Suppose, on the other hand, that $D(v) = D(a)$. Since the norm is strongly subdifferentiable at the partial isometry v, it follows from Lemma 6.3.3 that it is also strongly subdifferentiable at a. If follows that there is a partial isometry w so that $aw^* = ww^*$ with $\|a - w\| < 1$. By the argument given above, we must have $w = v$ which completes the proof of this part. (ii) Suppose $x \neq v$ and let $x = u|x|$ be the polar decomposition of x. As we have seen earlier, we must have $v = uv^*v$, so that $|x| \neq v^*v$ and we can find an element w in the C^*-algebra generated by $|x|$ with

$$v^*vw = v^*v, \quad \|v^*v - w\| < 1, \quad \||x| - w\| \geq 1.$$

(One can see how to do this by representing the algebra as a $C_0(Q)$.)

Since $v = uv^*v$, we have $uv^* = v^*v$ and by part (i), $v^*u = v^*v$. From this we get

$$v^*(uw) = v^*v, \quad \|v - uw\| = \|uv^*v - uw\| < 1,$$

and also

$$\|x - uw\| \geq \|u^*x - u^*uw\| = \||x| - w\| \geq 1.$$

This contradiction establishes the desired statement. □

We are now ready to give an alternate proof of Kadison's theorem. We will not restate the theorem again here, but recall its statement in Theorem 6.2.5.

PROOF. We first assume that $\mathcal{A}$ and $\mathcal{B}$ are von Neumann algebras. Let v be a partial isometry in $\mathcal{A}$. Since the norm is strongly subdifferentiable at 1, it must be subdifferentiable at any partial isometry by Lemma 6.3.5. Hence, because T is an isometry, the norm of $\mathcal{B}$ will be strongly subdifferentiable at $T(v)$. From the implication (i) implies (iii) of Theorem 6.3.6, there is a partial isometry $w \in \mathcal{B}$ with

$$T(v)w^* = ww^* \quad \text{and} \quad \|w - T(v)\| < 1.$$

For any $x \in F_{v,0}$, it is straightforward to see that

$$D(Tx) = (T^*)^{-1}D(x) = (T^*)^{-1}D(v) = D(Tv) = D(w).$$

It follows from Lemma 6.3.7(i) that $T(F_{v,0}) \subseteq F_{w,0}$, and the reverse containment can be shown similarly, so that we have $T(F_{v,0}) = F_{w,0}$. Now $\|T(v) - T(x)\| < 1$ for all $x \in F_{v,0}$, and by Lemma 6.3.7(ii), we obtain $Tv = w$. In fact, we have shown that $T(v)$ is a partial isometry and

$$T(F_{v,0}) = F_{T(v),0}$$

for every partial isometry $v \in \mathcal{A}$. If we let $F_v = \{a \in S(\mathcal{A}) : av^* = vv^*\}$, and $x \in F_v$, then the line segment $[x, v]$ lies in F_v as well. It is therefore possible to write $x = v + tw$ where $t \geq 1$ and $v + w$ lies in $F_{v,0}$. From the statement above, we must have $T(v + w) \in F_{T(v),0}$ and this can only happen if $T(w)T(v)^* = 0$. Hence, $T(x)T(v)^* = T(v)T(v)^*$, and $T(x) \in F_{T(v)}$. The opposite inclusion is shown in the same way. Thus,

$$T(F_v) = F_{T(v)}$$

for any partial isometry v.

Let Ψ be defined from $\mathcal{A}$ to $\mathcal{B}$ by

$$\Psi(a) = T(a)T(1)^*.$$

Given any projection $p \in \mathcal{A}$, we must have $1 \in F_p$ and so from what we established above,

$$T(1)T(p)^* = T(p)T(p)^* = [T(1)T(p)^*]^* = T(p)T(1)^* = \Psi(p).$$

A projection p is a partial isometry, and from above, $T(p)$ is also a partial isometry. Thus $T(p)T(p)^*$ is a projection which is different from zero, and it now follows from Lemma 6.3.1 that Ψ is a JB^*-isomorphism. It remains to show that $T(1)$ is unitary, from which it also follows that $\Psi(1) = 1$.

Choose $a \in \mathcal{A}$ such that $T(a) = 1 - T(1)^*T(1)$. Since $T(1)$ is a partial isometry, we see (as in earlier calculations) that

$$\Psi(a) = (1 - T(1)^*T(1))T(1)^* = 0,$$

and since Ψ is injective, we get $T(1)^*T(1) = 1$. We can apply the same arguments to $(T)^* : a \to T(a^*)^*$ to show that $T(1)T(1)^* = 1$, and so $T(1)$ is unitary.

We observe here that our expression for T is of the form

$$T = \Psi v$$

which is backward from what was advertised in the statement of the theorem originally. If we treat the associated isometry $(T)^*$ as defined just above, (note this is not the adjoint map), and get

$$(T)^* = \Psi v, \quad \text{where} \quad v = (T)^*(1),$$

we can define

$$\tau(a) = [\Psi(a^*)]^* \quad \text{and} \quad u = T(1)$$

to get $T = u\tau$ as desired.

To complete the proof we now suppose that $\mathcal{A}$ and $\mathcal{B}$ are unital C^*-algebras. The key is that the double duals, $\mathcal{A}^{**}$ and $\mathcal{B}^{**}$ are actually von Neumann algebras and the double adjoint T^{**} is an isometry. Hence, from the previous part we have a unitary $u \in \mathcal{B}^{**}$ and a JB^*-isomorphism τ_0 from $\mathcal{A}^{**}$ to $\mathcal{B}^{**}$ such that $T^{**} = u\tau_0$. Since $1 \in \mathcal{A}$ and $\tau_0(1) = 1$, $u = T(1) \in \mathcal{B}$, and we must have $\tau_0(\mathcal{A}) \subseteq \mathcal{B}$. If we let τ be the restriction of τ_0 to $\mathcal{A}$, we get $T = u\tau$ and the theorem is proved. $\square$

6.4. The Nonsurjective Case of Kadison's Theorem

Isometries on commutative C^*-algebras have been treated in Chapter 2 and the results in the nonsurjective case were given in Section 2.3, and specifically, Theorem 2.3.10. In the particular case where the isometry T mapped $C_0(Q)$ into $C_0(K)$ (Holsztynski's theorem), the form of T given by $Tf(t) = h(t)f(\varphi(t))$ only holds for t in a subset of K, (in fact, the Choquet boundary of the range of T), and how the functions Tf behave otherwise plays no role in the fact that T is an isometry. In this section we want to consider what happens to Kadison's theorem if we assume the operator from one C^*-algebra to another is not surjective. What behavior, if any, of T will be analogous to the restriction of range elements to a certain subset of their original domain?

In carrying out this investigation of the nonsurjective case, we are going to range a bit farther afield than usual, and in choosing the setting, touch base with one of the areas of functional analysis that is currently receiving much attention. This is the area of so-called *noncommutative functional analysis*, which features the study of operator spaces. Hence, it is *complete* isometries, rather than isometries, that will be of central interest. There is a very large body of literature that has arisen featuring a variety of approaches, and we will try to give an informal discussion that will lay enough groundwork to make the subsequent results understandable. In the notes we will suggest some references for the reader who wants to study the subject in detail.

Every Banach space can be embedded isometrically as a subspace of a commutative C^*-algebra $C(K)$, and the noncommutative generalizations naturally involve subspaces of the standard noncommutative C^*-algebra, the bounded operators on a Hilbert space. If $\mathcal{A}$ is a C^*-algebra, and M_n denotes the $n \times n$ complex matrices, then by $M_n(\mathcal{A})$ we will mean the set of $n \times n$ matrices with entries from $\mathcal{A}$. Note that $M_n(\mathcal{A})$ becomes a *-algebra in the natural way, and there is a unique way (since the norm is unique on a C^*-algebra) to introduce a norm such that $M_n(\mathcal{A})$ becomes a C^*-algebra. One way to see this is to represent $\mathcal{A}$ as a space of bounded operators on a Hilbert space H, and let $M_n(\mathcal{A})$ act on the direct sum of n copies of H in the obvious way.

A self-adjoint subspace $\mathcal{S}$ of a C^*-algebra $\mathcal{A}$ which contains 1 is called an *operator system*, while a subspace of $\mathcal{A}$ is called an *operator space*. For such an $\mathcal{S}$, we let $M_n(\mathcal{S})$ have the same norm and order structure that it inherits from $M_n(\mathcal{A})$. Given a linear map T from $\mathcal{S}$ to a C^*-algebra $\mathcal{B}$, and any positive integer n, we define an operator T_n from $M_n(\mathcal{S})$ to $M_n(\mathcal{B})$ by

$$T_n([a_{ij}]) = [T(a_{ij})].$$

We say that T is *completely bounded* if

$$\|T\|_{cb} = \sup_n \|T_n\| < \infty.$$

Furthermore, we say that T is *completely isometric* (*completely positive*) if T_n is isometric (positive) for each n. The transpose operator S, for example, on $M_2(\mathbb{C})$ is positive and has norm 1, but S_2 is not positive, and $\|S_2\| = 2$.

Our goal in this section is to try to characterize complete isometries from a C^*-algebra $\mathcal{A}$ into a C^*-algebra $\mathcal{B}$. It is a fact that the isometries between subspaces of $C(K)$ spaces are complete isometries, so that it is a natural generalization to consider complete isometries on the noncommutative C^*-algebras. Let us prove this last statement along with another observation that will be useful.

6.4.1. PROPOSITION. *Let $\mathcal{S}$ be a closed subspace contained in $C(Q)$ for some compact Q and which contains 1. Let ϕ be an isometry from $\mathcal{S}$ into*

$C_0(K)$ *for a locally compact Hausdorff space* K. *Then* ϕ *is a complete isometry.*

PROOF. For $t \in K$, let ϕ^t be the linear functional defined on $\mathcal{S}$ by $\phi^t(a) = \phi(a)(t)$. We want to show first that

$$\|\phi_n^t\| \leq \|\phi^t\|.$$

Let $A = [a_{ij}]$ be an element of $M_n(\mathcal{S})$ and let $x = (x_1, \dots, x_n)$ and $y = (y_1, \dots, y_n)$ be unit vectors in $\mathbb{C}^n$. Then

$$|\langle \phi_n^t([a_{ij}])x, y \rangle| = |\sum_{i,j} \phi^t(a_{ij})x_j\overline{y_i}| = |\phi^t(\sum_{i,j} a_{ij}x_j\overline{y_i})|$$

$$\leq \|\phi^t\| \cdot \|\sum_{i,j} a_{ij}x_j\overline{y_i}\|.$$

The element within the norm symbols on the last line above can be shown to be the entry in the $(1, 1)$ position of the product, in the C^*-algebra $M_n(C(Q))$, of matrices YAX where Y is the matrix of all zeros except the first row which consists of the elements $\overline{y_1} \cdot 1, \dots, \overline{y_n} \cdot 1$, and X is the matrix with all zeros except the first column which consists of the entries $x_1 \cdot 1, \dots, x_n \cdot 1$. Since $\|X\| = \|Y\| = 1$, we get

$$\|\sum_{i,j} a_{ij}x_j\overline{y_i}\| \leq \|YAX\| \leq \|A\|.$$

The conclusion from these last two inequalities is that

$$\|\phi_n^t(A)\| \leq \|\phi^t\|\|A\|,$$

which establishes our claim.

Now in $M_n(C_0(Q))$, an element G has norm given by $\sup_{t \in Q} \|G(t)\|$, and it can shown that

$$\|\phi_n\| = \sup\{\|\phi_n^t\| : t \in Q\} \leq \sup\{\|\phi^t\| : t \in Q\} = \|\phi\|.$$

This shows that ϕ is completely contractive, and the same argument applied to ϕ^{-1} shows that ϕ must be a complete isometry. $\square$

6.4.2. PROPOSITION. *If* ϕ *is an algebra* $*$-*homomorphism from a* C^*-*algebra* $\mathcal{A}$ *to a* C^*-*algebra* $\mathcal{B}$, *then* ϕ *is completely contractive. If* ϕ *is injective, then it is a complete isometry.*

PROOF. If a is a self-adjoint element of $\mathcal{A}$, then the nonzero elements of the spectrum of $\phi(a)$ are a subset of the nonzero elements of the spectrum of a; since the spectral radius of a self-adjoint element is equal to its norm, we have $\|\phi(a)\| \leq \|a\|$. Hence, given any element $a \in \mathcal{A}$, we see that

$$\|\phi(a)\|^2 = \|\phi(a)^*\phi(a)\| = \|\phi(a^*a)\| \leq \|a^*a\| = \|a\|^2.$$

Thus ϕ is contractive. The map ϕ_n will also be an algebra $*$-homomorphism on $M_n(\mathcal{A})$ to $M_n(\mathcal{B})$, and so it follows that ϕ_n is also contractive for each n.

Now if we knew that the range of ϕ was closed, and ϕ was injective, we could complete the proof by simply considering the inverse of ϕ. Unfortunately, we must work a bit harder. Let us assume that ϕ is one-to-one. We can see from the argument above that it is enough to show that ϕ is isometric on positive elements. Let us take $\mathcal{A}$ to be the (commutative) C^*-algebra generated by a fixed positive element, and take $\mathcal{B}$ to be the closure of $\phi(\mathcal{A})$. We may assume that units are adjoined to $\mathcal{A}$ and $\mathcal{B}$, and that $\phi(1) = 1$. Then $\mathcal{A}$ and $\mathcal{B}$ may be represented by continuous function spaces $C(Q), C(K)$ on compact Hausdorff spaces Q and K, respectively. The conjugate map of ϕ maps multiplicative linear functionals to multiplicative linear functionals, and so we may think of it as defining a map, which we call ϕ^*, that is continuous from K into Q. Since ϕ is an isomorphism and $\phi(\mathcal{A})$ is dense in $\mathcal{B}$, we have that ϕ^* is actually a bijection from K onto Q and, therefore, a homeomorphism. It follows that ϕ is an isometry, and thus an isometry on the original algebras as well.

Note that in the general case, if I is the kernel of ϕ, then I is a closed ideal in $\mathcal{A}$ and ϕ induces an isomorphism from the C^*-algebra $\mathcal{A}/I$ into $\mathcal{B}$ with image $\phi(\mathcal{A})$. By the argument in the previous paragraph, this isomorphism must be an isometry, and we conclude that the range of ϕ is closed.

Since the above arguments apply to ϕ_n for each n, we see that ϕ must be a complete isometry whenever it is injective.

$\square$

The fact that a $*$-isomorphism between C^*-algebras is an isometry is, of course, well known to those familiar with C^*-algebra theory.

We want to prove another result similar to that of Proposition 6.4.2, but in a more general setting. We will define a *triple system* to be a uniformly closed subspace X of a C^*-algebra such that $XX^*X \subseteq X$. A bounded linear map Φ between two triple systems $\mathcal{A}$ and $\mathcal{B}$ is called a J^*-*isomorphism* if it is a bijection which satisfies

$$(115) \qquad \Phi(aa^*a) = \Phi(a)\Phi(a)^*\Phi(a)$$

for all $a \in \mathcal{A}$. An operator on a triple system is called a *triple morphism* if it is bounded, linear, and satisfies

$$(116) \qquad \Phi(xy^*z) = \Phi(x)\Phi(y)^*\Phi(z).$$

We warn the reader that language and notation may vary in regard to these notions.

6.4.3. PROPOSITION. *A one-to-one triple morphism Φ between two triple systems $\mathcal{A}$ and $\mathcal{B}$ is completely isometric.*

PROOF. Recall from spectral theory that if b is a self-adjoint element in a C^*-algebra, then $\|b^3\| = \|b\|^3$. Hence, for any element a of a triple system,

$$\|aa^*a\|^2 = \|(aa^*a)^*(aa^*a)\| = \|(a^*a)^3\| = \|a^*a\|^3 = \|a\|^6.$$

Therefore, we have $\|aa^*a\| = \|a\|^3$, and if we apply (115), we obtain

$$\|\Phi(a)\|^3 = \|\Phi(a)\Phi(a)^*\Phi(a)\| = \|\Phi(aa^*a)\| \leq \|\Phi\|\|a\|^3.$$

We must, therefore, have $\|\Phi\| \leq 1$. It is straightforward to see that for any positive integer n, $M_n(X)$ is a triple system and Φ_n is a triple morphism between two triple systems. Thus Φ_n is contractive, and Φ is a completely contractive. The range of a triple morphism is necessarily closed, and since Φ is injective, by the closed graph theorem and the previous argument applied to Φ^{-1} we conclude that Φ is a complete isometry.   □

The fact that Φ, in the previous proposition, is contractive would hold even if Φ is only bounded and satisfies (115). Indeed, it can be seen from this that a J^*-isomorphism must be an isometry, but there is a difficulty in trying to extend the result to the matrix spaces.

Consider the operator $T : C([0,1] \to C([0,1])$ defined by

$$Tf(t) = \frac{1}{2}(f \circ \varphi_1)(t) + \frac{1}{2}(f \circ \varphi_2)(t),$$

where φ_1 and φ_2 are continuous functions on $[0,1]$ to itself such that $\Gamma = \{t \in [0,1] : \varphi_1(t) = \varphi_2(t)\}$ is a proper closed subinterval of $[0,1]$, and $\varphi_1(\Gamma) = [0,1]$. (Recall Example 2.3.17 from Chapter 2.) If we think of the range of T as being in the second dual space of $C([0,1])$ and take p to be a function defined on $[0,1]$ which is 1 on elements of Γ and 0 otherwise, then the operator $R = p\widehat{T}(\cdot)p$ is called a *reducing $\wedge$-compression* of T. Note that R is an isometry from $C([0,1])$ into $C([0,1])^{**}$ and if $S = (1-p)\widehat{T}(\cdot)(1-p)$, then $T = R + S$, and $\|T(f)\| = max\{\|R(f)\|, \|S(f)\|\}$ for all $f \in C([0,1])$. Observe that the expression for R can be given by

$$R(f) = f \circ \varphi_1 = u\theta(f)$$

where u is a partial isometry (in this case, just the identity) and θ is $*$-homomorphism defined on $C([0,1])$. Furthermore, it is this part which makes T a complete isometry. Of course our example is in the commutative case, but it is this same kind of behavior which will characterize complete isometries in the general case.

If T is a map from an operator space X into a von Neumann algebra $\mathcal{M}$, we say that R is a *reducing compression* of T if there exist projections $p, q \in \mathcal{M}$ such that $R = qT(\cdot)p = T(\cdot)p = qT(\cdot)$. Letting $S = (1-q)T(\cdot)(1-p)$, we have $R + S = T$, and $\|T(x)\| = max\{\|R(x)\|, \|S(x)\|\}$ for $x \in X$ and similarly for matrices. (If one represents $\mathcal{M}$ as operators on a Hilbert space, orthonormal bases may be chosen in such a way as to write $T(\cdot)$ as a block diagonal matrix with R and S on the diagonal. It is this idea which gives motivation to the general definition.) In case $T : X \to \mathcal{B}$, where $\mathcal{B}$ is a C^*-algebra, we say that $R : X \to \mathcal{B}^{**}$ is a reducing $\wedge$-compression of T if R is a reducing compression of $\widehat{T}$. Here, $\widehat{T}$ is the composition of T followed by the canonical embedding of $\mathcal{B}$ into the von Neumann algebra $\mathcal{B}^{**}$.

One of the keys to the proof of our main theorem is the notion of a *triple envelope* of an operator space. If X is such a space, there exists a pair (Z, j) consisting of a triple system Z and a linear complete isometry $j : X \to Z$ whose range generates Z as a triple system, and which has the following universal property: for any other pair (W, ϕ) consisting of a triple system W and a complete isometry ϕ whose range generates W as a triple system, there exists a (necessarily unique and surjective) triple morphism $\pi : W \to Z$ such that $\pi \circ \phi = j$.

We need one more piece of terminology. If J is a left ideal in a C^*-algebra $\mathcal{A}$, then $J^{\perp\perp}$ is a weak* closed left ideal in the von Neumann algebra $\mathcal{A}^{**}$. Thus, there exists a projection $p \in \mathcal{A}^{**}$ with $J^{\perp\perp} = \mathcal{A}^{**}p$, and

$$(\mathcal{A}/J)^{**} \cong \mathcal{A}^{**}/J^{\perp\perp} \cong \mathcal{A}^{**}(1-p)$$

completely isometrically. The projection p may be taken to be any weak* limit point in $\mathcal{A}^{**}$ of a right contractive approximate identity for J. The projection p is called the (right) *support projection* for J. If J is a two-sided ideal then p is a projection in the center of $\mathcal{A}^{**}$.

Here is the main theorem. Our proof will be a bit sketchy.

6.4.4. THEOREM. *(Blecher and Hay) Let $T : \mathcal{A} \to \mathcal{B}$ be a completely contractive linear map between C^*-algebras. The following are equivalent:*

(i) *T is a complete isometry,*

(ii) *T possesses a reducing $\wedge$-compression which is a $1-1$ triple morphism,*

(iii) *there exists a C^*-subalgebra $\mathcal{D}$ of $\mathcal{B}$, a closed two-sided ideal J in $\mathcal{D}$, a $*$-isomorphism $\pi : \mathcal{A} \to \mathcal{D}/J$, and a partial isometry $u \in \mathcal{B}^{**}$, such that*

$$\widehat{T}(a)(1-p) = u\pi(a)$$

and such that $u^*\widehat{T}(a) = \pi(a)$, for all $a \in \mathcal{A}$ where p is the support projection for J in $\mathcal{D}^{**}$, viewed as an element of $\mathcal{B}^{**}$. We are viewing $\pi(\mathcal{A})$ in the equations above as a subset of $(1-p)\mathcal{B}^{**}(1-p)$ via the identifications $\mathcal{D}/J \subset \mathcal{D}^{**}/J^{**} = (1-p)\mathcal{D}^{**}(1-p) \subset (1-p)\mathcal{B}^{**}(1-p)$,

(iv) *there is a projection p in $\mathcal{B}^{**}$, a partial isometry u in $\mathcal{B}^{**}(1-p)$, and a 1-1$*$-homomorphism $\theta : \mathcal{A} \to (1-p)\mathcal{B}^{**}(1-p)$, such that*

$$\widehat{T(\cdot)}(1-p) = u\theta(\cdot)$$

and such that $u^*\widehat{T(\cdot)} = \theta(\cdot)$.

PROOF. A 1-1 triple morphism satisfies (116), and is a complete isometry by Proposition 6.4.3. From this we see that (ii) implies (i). Given that (iii) is satisfied and by Proposition 6.4.2, for $a \in \mathcal{A}$ we have

$$\|T(a)\| \geq \|\widehat{T}(a)(1-p)\| = \|u\pi(a)\| \geq \|u^*u\pi(a)\| = \|\pi(a)\| = \|a\|.$$

A similar inequality would hold for the matrix case, and since T is completely contractive, we see that it must be a complete isometry. The argument for (iv) implies (i) is exactly the same.

Suppose now that T is a complete isometry.

For (ii) let Z be the triple subsystem of $\mathcal{B}$ generated by $T(\mathcal{A})$. Since T is a complete isometry, and since $\mathcal{A}$ is its own triple envelope, there is, by the property of triple envelopes, a surjective triple morphism $\mu : Z \to \mathcal{A}$ such that

$$\mu(T(a)) = a$$

for all $A \in \mathcal{A}$. If N denotes the kernel of the mapping μ, then Z/N is a triple system and there is a triple morphism $\gamma : \mathcal{A} \to Z/N$ defined by $\gamma(a) = Ta + N$. If we let p and q denote the right and left support projections of N, we can think of p as an element in $(Z^*Z)^{**} \subset \mathcal{B}^{**}$. Furthermore we have

$$\hat{z}p = q\hat{z} = q\hat{z}p$$

for all $z \in Z$. Consider the map $\rho : Z/N \to (1-q)\mathcal{B}^{**}(1-p)$ where

$$\rho(z + N) = \hat{z}(1-p).$$

If $\rho(z + N) = 0$, then $\hat{z} = \hat{z}p$ so

$$\hat{z} \in \hat{Z} \cap Z^{**}p = \hat{N}.$$

Therefore, $z \in N$, and ρ is one-to-one. It is straightforward to show that

$$\widehat{xy^*z}(1-p) = \hat{x}(1-p)(\hat{y}(1-p))^*\hat{z}(1-p)$$

which means that ρ is a 1-1 triple morphism and therefore a complete isometry. Thus $\rho \circ \gamma$ is a 1-1 triple morphism on $\mathcal{A}$ into $(1-q)\mathcal{B}^{**}(1-p)$ given by $\widehat{Ta}(1-p) = (1-q)\widehat{Ta}(1-p) = (1-q)\widehat{Ta}$. This gives a reducing $\wedge$-compression of T which is a triple morphism.

For (iii), consider again the triple isomorphism $\gamma : \mathcal{A} \to Z/N$ from the previous paragraph. We may view Z/N as a triple subsystem $W = \hat{Z}(1-p)$ of $\mathcal{B}^{**}(1-p)$ by means of the map $\rho(z + N) = \hat{z}(1-p)$. Since p commutes with the image of Z^*Z in $\mathcal{B}^{**}$, we have

$$T(a_1)^*T(a_2)(1-p) = (1-p)T(a_1)^*T(a_2) = (1-p)T(a_1)^*T(a_2)(1-p)$$

for all $a_1, a_2 \in \mathcal{A}$. If we define $\theta(a_1^*a_2) = \rho(a_1)^*\rho(a_2)$, then $\theta : \mathcal{A} \to W^*W$ can be shown to be a 1-1, surjective $*$-homomorphism because ρ is a 1-1 triple morphism. Choose u to be any weak* limit point of $\{\widehat{Te_\alpha}(1-p)\}$ where $\{e_\alpha\}$ is any contractive approximate identity in $\mathcal{A}$. Then

$$u\theta(a_1^*a_2) = \lim \rho(e_\alpha)\rho(a_1)^*\rho(a_2) = \lim \rho(e_\alpha a_1^*a_2) = \rho(a_1^*a_2).$$

This gives $\rho = u\theta(\cdot)$, and in a similar way we can show that $u^*\rho(a) = \theta(a)$. Hence, $u^*ub = b$ for all b in the range of θ and this will extend to all b in the weak* closure of the Range of θ. Thus u is a partial isometry and we have

$$\widehat{T(a)}(1-p) = u\theta(a) \quad \text{and} \quad u^*u\theta(a) = \theta(a)$$

for all $a \in \mathcal{A}$. If we let $D = Z^*Z$, and $J = N^*N$, we get that $W^*W \cong D/J$. Write π for the $*$-isomorphism corresponding to θ under this equivalence. This gives (iii) and (iv) as well. $\square$

We remark that in case T is unital, the space Z in the proofs can be taken to be a C^*-algebra itself, and the triple morphisms become *-homomorphisms. Also, the u in (iii) and (iv) can be taken to be $T(1)$.

In the commutative case, we recall that an isometry must be completely isometric, and since closed ideals of $C_0(K)$ correspond to sets of functions vanishing on some subset of K, it is possible to essentially recover Holsztyn-ski's characterization of nonsurjective isometries on $C_0(K)$ spaces from the theorem above.

6.5. The Algebras $C^{(1)}$ and AC

We return here to function spaces which are also algebras. Howeve,r the algebras under consideration will not be uniform algebras. The goal is to show that an isometry on these spaces determines an algebra isomorphism. The spaces we wish to study are

(i) the algebra $C^{(1)}([0,1])$ of complex functions continuously differentiable on $[0,1]$, with norm given by

(117) $$\|f\| = \max_{t \in [0,1]} (|f(t)| + |f'(t)|) \text{ for } f \in C^{(1)}$$

and

(ii) the algebra $AC([0,1])$ of absolutely continuous complex functions on $[0,1]$, with norm

(118) $$\|f\| = \max_{t \in [0,1]} |f(t)| + \int_0^1 |f'(s)| ds \text{ for } f \in AC.$$

We are going to show that an isometry on $C^{(1)}$ (and on AC as well) actually has the canonical form. We will suppress writing the interval $[0,1]$ when referring to the spaces.

First we show that for every $t \in [0,1]$, there is a function in $C^{(1)}$ which is a "peak" function in the sense that it assumes the max in equation (117) at t and only at t.

6.5.1. PROPOSITION. *For each* $t \in [0,1]$ *and* $\theta \in [-\pi, \pi]$, *there exists* $h \in C^{(1)}$ *such that*

$$|h(t)| + |h'(t)| > |h(s)| + h'(s)|$$

for all $s \in [0,1]$ *with* $s \neq t$. *Furthermore,* $|h(t)| = h(t) > 0$ *and* $|h'(t)| = e^{i\theta} h'(t) > 0$.

PROOF. We start by choosing a real nonnegative continuous function f for which $f(t) = 1$ and where the slope of f is 1 on $(0, t)$ (if this set is nonempty) and -1 on $(t, 1)$ (if this set is nonempty). Define $g \in C^{(1)}$ by

$$g(s) = \int_0^s f(u) du - \int_0^t f(u) du.$$

It is straightforward to show now that $g(t) = 0$ and

$$|g(t)| + |g'(t)| > |g(s)| + |g'(s)|$$

for all $s \in [0,1]$ with $s \neq t$. The function

$$h(s) = e^{-i\theta} g(s) + 1$$

has the properties claimed in the statement of the proposition.   □

We now associate $C^{(1)}$ with a subspace of a $C(K)$ space. Let $K = [0,1] \times [-\pi, \pi]$, and for each $f \in C^{(1)}$ let $R(f) = \tilde{f}$ be defined on K by

$$\tilde{f}(t, \theta) = f(t) + e^{i\theta} f'(t).$$

Then R is a linear isometry from $C^{(1)}$ onto the closed subspace $N = \{\tilde{f} : f \in C^{(1)}\}$ of $C(K)$. It is clear that $\|R(f)\| \leq \|f\|$ and the reverse inequality follows since for a given t, there is some θ such that $|f(t) + e^{i\theta} f'(t)| = |f(t)| + |f'(t)|$.

Ultimately, we are going to make use of the extreme point method, and to determine extreme points of the unit ball of $C^{(1)}$, we can make use of our knowledge of the form of the extreme points in $B(N^*)$. Given $(t, \theta) \in K$, we let $\psi(t, \theta)$ denote the bounded linear functional on $C^{(1)}$ given by

$$\psi(t, \theta)(f) = f(t) + e^{i\theta} f'(t).$$

It is probably not surprising that these functionals determine the extreme points of $B((C^{(1)})^*)$.

6.5.2. LEMMA. *The extreme points of the unit ball of $(C^{(1)})^*$ are precisely the functionals of the form $e^{i\gamma} \psi(t, \theta)$ for some $\gamma \in [-\pi, \pi]$ and $(t, \theta) \in K$.*

PROOF. If $x^* \in ext((C^{(1)})^*)$, then $(R^{-1})^* x^*$ is an extreme point in the dual unit ball of $C(K)$. Hence there exists $(t, \theta) \in K$ such that

$$(R^{-1})^*(x^*)(Rf) = e^{i\gamma} Rf(t, \theta)$$

for some $\gamma \in [-\pi, \pi]$. But it is clear that $e^{i\gamma} Rf(t, \theta) = e^{i\gamma} \psi(t, \theta)(f)$. Hence x^* has the desired form.

On the other hand, let $t \in [0, 1]$ and $\theta \in [-\pi, \pi]$ be given. Let h be the function guaranteed by Proposition 6.5.1. Then $\tilde{h}$ "peaks" at $(t, \theta) \in K$ in the sense that $\|\tilde{h}\| = \tilde{h}(t, \theta) > |\tilde{h}(s, \phi)|$ for any $(s, \phi) \in K$ different from (t, θ). In any Banach space, given any element, there is an extreme point in the dual unit ball whose value at the element is its norm. The evaluation functional (which using our previous notation would be denoted by $\psi_{(t,\theta)}$) corresponding to (t, θ) is the only evaluation functional which satisfies the requirement mentioned above for $\tilde{h} \in N$ and so it must be an extreme point for the unit ball of N^*. Hence $\psi(t, \theta) = R^*(\psi_{(t,\theta)}$ must be an extreme point for the unit ball of $(C^{(1)})^*$.   □

Our goal is to show that any surjective isometry on $C^{(1)}$ is a unimodular multiple of a composition operator, that is, of the form $f \to e^{i\lambda} f \circ \varphi$, where φ is defined on $[0, 1]$. Hence we can expect $T(1)$ to be a constant. Since we will need that fact to help us prove the theorem, let us make that observation now.

It is clear that any extreme point of $B((C^{(1)})^*)$ has absolute value 1 on the 1 function. Hence, if T is an isometry from $C^{(1)}$ onto itself, then we must have

$$|e^{i\gamma}\psi_{(,\theta)}(T(1))| = |e^{i\gamma}\psi(t,\theta)(1)| = 1$$

for all $(t,\theta) \in K$. Since $\|T(1)\| = 1$, for each $\in [0,1]$ it is true that

$$1 \geq |T(1)(t)| + |T(1)'(t)| \geq |T(1)(t) + e^{i\theta}T(1)'(t)| = 1$$

for each $\theta \in [-\pi,\pi]$. Thus $|T(1)'(t)|$ must be either 1 or 0 for each t and by continuity it is necessarily identically 0. This yields the following lemma.

6.5.3. LEMMA. *If T is a surjective linear isometry on $C^{(1)}$, then*

$$T(1)(t) \equiv e^{i\lambda}$$

for some $\lambda \in [-\pi,\pi]$.

Our previous experience with the extreme point method leads us to expect to define the function φ by utilizing a pairing between extreme points of the dual ball. However, extreme points involve a pair (t,θ) from $[0,1] \times [1\pi,\pi]$, while φ is to be a function of t alone. If T is an isometry, and $\psi(t,\theta) \in ext((C^{(1)})^*)$, then $T^*(\psi(t,\theta) = e^{i\gamma}\psi(s_{(t,\theta)},\phi_{(t,\theta)})$. We want to show that $s_{(t,\theta)}$ is a function of t only.

6.5.4. LEMMA. *In the notation introduced just above, for each $t \in [0,1]$, $s_{(t,\theta)} = s_{(t,0)}$ for each $\theta \in [-\pi,\pi]$.*

PROOF. Let $t \in [0,1]$, and define a map $\tau : [-\pi,\pi] \to [0,1]$ by

$$\tau(\theta) = s_{(t,\theta)}.$$

Suppose τ is not continuous at θ. Then we may assume there is a sequence $\{\theta_n\}$ and a neighborhood V of $s_{(t,\theta)}$ so that $\theta_n \to \theta$ but $\tau(\theta_n) \notin V$ for every n. Let h be the element of $C^{(1)}$ corresponding to the pair $(s_{(t,\theta)},\phi_{(t,\theta)})$ as given in Proposition 6.5.1. Since $[0,1]\backslash V$ is compact, the continuous function $|h(s)| + |h'(s)|$ assumes a maximum α_0 on the compact set which is necessarily strictly less than the norm of h. Since $(t,\theta_n) \to (t,\theta)$ in K, we must have $\psi(t,\theta_n) \to \psi(t,\theta)$ in the weak*-topology, from which we may conclude

$$|h(s_{(t,\theta_n)} + e^{i\phi(t,\theta_n)}h'(s_{(t,\theta_n)})| \to |h(s_{(t,\theta)}) + e^{i\phi(t,\theta)}h'(s_{(t,\theta)})| = \|h\|.$$

This leads to a contradiction.

The image of $[-\pi,\pi]$ under the continuous function τ must be a closed subinterval of $[0,1]$, and, in fact, it must be a singleton. Suppose there is a nondegenerate subinterval of this interval called J, and there is some γ so that $s_{(t,\gamma)}$ is not in J. Choose $g \in C^{(1)}$ such that g is identically zero on J and $|g'(s_{(t,\gamma)}| > |g(s_{(t,\gamma)}| > 0$. Then we see that

$$Tg(t) + e^{i\theta}(Tg)'(t) = 0 \text{ for infinitely many } \theta,$$

and

$$Tg(t) + e^{i\gamma}(Tg)'(t) \neq 0.$$

This simply cannot hold. Hence the range of τ is a singleton, which means that $s_{(t,\theta)} = s_{(t,0)}$ for each θ. $\qquad\square$

Because of what we have just proved, we can define a function φ from $[0,1]$ to itself by letting

$$\varphi(t) = s_{(t,0)}.$$

By also considering T^{-1}, it is easy to see that φ is a bijection. We are ready to prove the main theorem.

6.5.5. THEOREM. *(Cambern) Let T be an isometry from $C^{(1)}$ onto itself. Then there is a function φ which is a homeomorphism on $[0,1]$ and $\lambda \in [-\pi, \pi]$ such that*

$$Tf(t) = e^{i\lambda} f(\varphi(t)),$$

for all $f \in C^{(1)}$ and $t \in [0,1]$. Furthermore, φ is one of the functions j or $1-j$ where j is the identity mapping $j(x) = x$.

PROOF. Let $t \in [0,1]$ and consider the function g defined in the proof of Proposition 6.5.1 which has the property that $g(t) = 0$, $g'(t) > |g(s)| + |g'(s)|$ for all s in $[0,1]$ with $s \neq t$. Given any θ in $[-\pi, \pi]$, we must have

$$\|g\| = e^{-i\theta} \psi(t, \theta)(g) = e^{-i\theta} T^* \psi(t, \theta)(T^{-1}(g)) = e^{i(\lambda-\theta)} \psi(\varphi(t), \phi_{(t,\theta)})(T^{-1}g),$$

where we have used the constant value of $T(1)$ from Lemma 6.5.3. For this equation to hold for all θ, it is necessary that $(T^{-1}g)(\varphi(t)) = 0$ and also $\phi_{(t,\theta)} = \phi_{(t,0)} + \theta$.

Now suppose $f \in C^{(1)}$ with $f(t) = 0$, so that for all $\theta \in [-\pi, \pi]$ we obtain

$$f'(t) = e^{-i\theta} \psi(t, \theta)(f) = e^{-i\theta} T^* \psi(t, \theta)(T^{-1}(f))$$
$$= e^{i(\lambda-\theta)} \psi(\varphi(t), \phi_{(t,0)+\theta})(T^{-1}(f)).$$

Again we must conclude that $(T^{-1}(f))(\varphi(t)) = 0$.

Given an arbitrary f in $C^{(1)}$, we define g by $g(s) = f(s) - f(t)$. Then $g(t) = 0$ and by what we saw above,

$$0 = T^{-1}g(\varphi(t)) = T^{-1}f(\varphi(t)) - f(t)(T^{-1}1)(\varphi(t))$$
$$= T^{-1}f(\varphi(t)) - e^{-i\lambda} f(t).$$

Upon replacing f by Tf, we see that

$$Tf(t) = e^{i\lambda} f(\varphi(t)).$$

For $f = j$, the identity mapping, the above equation leads to

$$\varphi(t) = e^{-i\lambda}(Tj)(t),$$

from which we conclude that φ is an element of $C^{(1)}$. Hence it is a homeomorphism, and from the form of the isometry T, the function φ must satisfy $|\varphi'(t)| \equiv 1$. It is therefore clear that $\varphi = j$ or $\varphi = 1 - j$. $\qquad\square$

We will state the theorem for the algebra AC and indicate the broad steps in the proof, but we omit the details.

6.5.6. THEOREM. *(Cambern) Let T be an isometry from $AC([0,1])$ onto itself. Then*

$$Tf(t) = e^{i\lambda} f(\varphi(t))$$

for all $f \in AC$ and $t \in [0,1]$, and with $e^{i\lambda} = T(1)$. The function φ is given by $\varphi = e^{-i\lambda} T(j)$, where j is the identity mapping of $[0,1]$ onto itself.

Let V denote the closed unit ball of the space $L^\infty([0,1])$ which is compact in the weak*-topology. Let K denote the compact space $[0,1] \times V$ and for $f \in AC$, define $\tilde{f} \in C(K)$ by

$$\tilde{f}(t,v) = f(t) + \int_0^1 f'(s)\overline{v}(s)ds.$$

Then $f \to \tilde{f}$ establishes an isometry between AC and a closed subspace N of $C(K)$. For $(t,v) \in K$, the equation

$$\psi(t,v)(f) = f(t) + \int_0^1 f'(s)\overline{v}(s)ds$$

defines a bounded linear functional on AC. As we saw in the discussion of $C^{(1)}$, the extreme points of the unit ball of $(AC)^*$ constitute a subset of the collection of functionals $e^{i\gamma}\psi(t,v)$. If $\psi(t,v)$ is extreme, then v is an extreme point of the unit ball of L^∞ and so has absolute value 1 almost everywhere in $[0,1]$.

For a given t in $[0,1]$, let v_t denote the L^∞ function which is 1 on $[0,t)$ (if nonvoid) and -1 on $(t,1]$ (if nonvoid). Then for all $t \in [0,1]$ and $\theta \in (-\pi/2, \pi/2)$, the functional $\psi(t, e^{i\theta}v_t)$ is an extreme point for the unit ball of $(AC)^*$. As before, it is shown that for an isometry T, $T(1)$ is a unimodular constant.

For $(t,\theta) \in [0,1] \times (-\pi/2, \pi/2)$, we have

$$T^*\psi(t, e^{i\theta}v_t) = e^{i\gamma}\psi(s_{(t,\theta)}, \beta_{(t,\theta)}).$$

The key is to show that

$$\beta_{(t,\theta)} = e^{i\theta}\beta_{(t,0)} \text{ and } s_{(t,\theta)} = s_{(t,0)}$$

for all θ. The function φ can be defined by

$$\varphi(t) = s_{(t,0)}$$

and the theorem is proved in much the same way as before.

6.6. Douglas Algebras

A *Douglas algebra* is a closed subalgebra of $L^\infty(\mathbf{T})$ which contains $H^\infty(\mathbf{T})$ where $\mathbf{T}$ is the unit circle endowed with Lebesgue measure. By the Gelfand theory, we may think of L^∞ as a space $C(K)$ of continuous functions on a compact Hausdorff space K, which is the maximal ideal space, or equivalently, the set of multiplicative linear functionals on L^∞ endowed with the weak*-topology. Thus H^∞ can be regarded as a closed subspace of $C(K)$ which contains the constants, and is a strongly separating function subspace of L^∞, in the language introduced in Chapter 2. Hence any Douglas algebra has the same properties. If T is any linear isometry from one Douglas algebra onto another, then it follows from the de Leuuw, Rudin, Wermer Theorem (2.3.16) that T is a unimodular multiple of an algebra isomorphism between the two algebras.

We want to be a bit more precise than that. If we make use of Theorem 2.5.3, we see that for a surjective isometry $T : M \rightarrow N$, where M, N are Douglas algebras, there is a homeomorphism φ between the Šilov boundaries, ∂N and ∂M, and a unimodular function h in N so that

$$Tf(t) = h(t)f(\varphi(t)) \text{ for all } t \in \partial N.$$

It is known that the Šilov boundary of H^∞ is equal to the maximal ideal space for L^∞, that is, our set K. Hence, the homeomorphism φ is defined on all of K, and the map $Sf = f \circ \varphi$ defines an algebra isomorphism on all of $C(K)$, or by identification under the Gelfand theory, all of L^∞. We now prove a theorem describing such algebra isomorphisms.

6.6.1. THEOREM. *(Font) Let S be an algebra isomorphism of L^∞ onto itself. Then $Sf = f \circ Sj$ for all $f \in L^\infty$, where j denotes the identity function $j(z) = z$ on the circle.*

PROOF. It is well known that an algebra homomorphism from a commutative semisimple Banach algebra to another is continuous. Thus S is bounded and has $\|S\| \geq 1$. We first want to show that for a closed subset U of $\mathbf{T}$, we necessarily have

(119) $$S\chi_U = \chi_U \circ Sj.$$

Suppose $(S\chi_U)(z) = 1$ for some $z \in \mathbf{T}$. For $\epsilon > 0$, we let $\{U_1, U_2, \ldots, U_n\}$ denote a partition of T into arcs with lengths each less than $(1\epsilon/\|S\|)$, and for each i let α_i be the "midpoint" of the arc U_i. Also, for each i, let $V_i = U_i \cap U$ and $W_i \cap (\mathbf{T}\backslash U)$. Then

$$\left\| j - \sum_{i=1}^n \alpha_i \chi_{V_i} - \sum_{l=1}^n \alpha_l \chi_{W_l} \right\| < \frac{\epsilon}{\|S\|},$$

and from the continuity of S, it follows that

$$\left\| Sj - \sum_1^n \alpha_i S\chi_{V_i} - \sum_l^n \alpha_l S\chi_{W_l} \right\| < \epsilon.$$

We note that $S\chi_U$ is a characteristic function, since $S(\chi_U^2) = (S\chi_U)^2$. Thus $S\chi_U$ takes on only the values $-1, 0, 1$, and if $S\chi_U$ takes on the value -1 on a set of positive measure, then $S\chi_{(\mathbf{T}\backslash U)} = 2$ on that subset, which is a contradiction. By the cozero set of a function f we will mean the set $coz(f)$ of $s \in \mathbf{T}$ such that $f(s) \neq 0$. The cozero sets of the functions $S\chi_{V_i}$ are pairwise disjoint, and so $(S(\chi_{V_i})(z) = 1$ for exactly one $i = i_0$, while the values for other choices of i are zero. Hence, by the inequality displayed above, we must conclude that $|Sj(z) - \alpha_{i_0}| < \epsilon$. It now follows that the distance from $Sj(z)$ to U is less than ϵ, and since U is closed, $Sj(z) \in U$, and $(\chi_U \circ Sj)(z) = 1$.

Now $S\chi_U$ is a characteristic function, and so has only the values 1 and 0. If $S\chi_U(z) = 0$, we need to show that $\chi_U(Sj(z)) = 0$. Suppose that is not the case, and there exists a subset W of $\mathbf{T}$ with positive measure so that for all $s \in W$, $S\chi_U(s) = 0$, and $Sj(s) \in U$. Now we must have $coz(S^{-1}\chi_W) \cap U = \emptyset$. Since $coz(S^{-1}\chi_W)$ has positive measure, it contains a compact subset Q of positive measure. There is a set L so that $S^{-1}\chi_W = \chi_L$ from which we have $Q \subset L$. Then $coz(S\chi_Q) \subset coz(S\chi_L) = coz(S\chi_W) = W$. (Here we have used the fact that, in general, if $A \subset B$, then $coz(S\chi_A) \subset coz(S\chi_B)$.) If $s \in coz(S\chi_Q)$, then $(Sj)(s) \in Q$ as we saw in the previous paragraph. However, $s \in W$ and by the definition of W, $Sj(s) \in U$, which contradicts the fact that $Q \cap U = \emptyset$. This completes the proof of (119).

We should observe also that the range of Sj is contained in $\mathbf{T}$. Given $z \in \mathbf{T}$, let W be a closed set containing z. Then $\chi_W = S\chi_U$ for some closed set U. Thus $(S\chi_U)(z) = 1$, and by what we have above, $Sj(z) \in U \subset \mathbf{T}$.

To prove the theorem, let us suppose there is some $f \in L^\infty$ with $Sf \neq f \circ Sj$. This means that there is some $\epsilon > 0$ and a set W of positive measure in $\mathbf{T}$ such that

$$|(Sf - f \circ Sj)(z)| > \epsilon \text{ for all } z \in W.$$

By the definition of the norm in L^∞ there is a complex number α and a compact subset V of $coz(S^{-1}\chi_W)$ with positive measure such that

$$\|f\chi_V - \alpha\chi_V\| < \frac{\epsilon}{4\|S\|}.$$

As before, we get that $coz(S\chi_V) \subset W$, and so for every $z \in coz(S\chi_V)$ it is true that $S\chi_V(z) = 1$ and

$$\epsilon < |(Sf)(z)(S\chi_V) - (f \circ Sj)(z)(S\chi_V(z)|$$
$$= |(S(f\chi_V))(z) - (f \circ Sj)(z)(\chi_V \circ Sj)(z)|$$

for almost all $z \in coz(S\chi_V)$. In the equality we have used the multiplicativity of S as well as (119). We can write

$$|S(f\chi_V))(z) - ((f\chi_V \circ Sj)(z)| > \epsilon$$

for almost all $z \in coz(S\chi_V)$.

We now reach a contradiction, because we also have

$$\|S(f\chi_V) - (f\chi_V) \circ Sj\| \leq \|S(f\chi_V) - S(\alpha\chi_V)\| + \|S(\alpha\chi_V) - (f\chi_V) \circ Sj\|$$
$$\leq \frac{\epsilon}{4} + \|\alpha\chi_V \circ Sj - (f\chi_V) \circ Sj\|$$
$$= \frac{\epsilon}{4} + \|(\alpha\chi_V - f\chi_V) \circ Sj\|$$
$$\leq \frac{\epsilon}{4} + \|\alpha\chi_V - f\chi_V\| < \frac{\epsilon}{2}.$$

$\square$

Let us summarize what we have done in the following theorem.

6.6.2. THEOREM. *(Font) Let $\mathcal{A}$ and $\mathcal{B}$ be Douglas algebras and suppose T is a linear isometry from $\mathcal{A}$ onto $\mathcal{B}$. Then there is a function $h \in \mathcal{B}$ which is unimodular and an algebra isomorphism S of L^∞ onto itself such that $Tf = h(f \circ Sj)$, where $j(z) = z$ for all z. Furthermore, the inverse of Sj is $S^{-1}j$ almost everywhere.*

PROOF. Our discussion prior to the statement of Theorem 6.6.1 shows how to get h and S. Note that $h = T1$, so $\|h\| = 1$. Furthermore, there exists $f \in \mathcal{A}$ with $\|f\| = 1$ and such that $Tf = 1$. Hence, for any $z \in \mathbf{T}$, we have

$$1 = |Tf(z)| = |h(z)||(f \circ Sj)(z)| \leq |h(z)| \leq 1.$$

To complete the proof, observe that for any $g \in \mathcal{B}$ we have

$$S^{-1}g = g \circ S^{-1}j$$

from which we get

$$j = S^{-1}(Sj) = Sj \circ S^{-1}j \text{ everywhere.}$$

$\square$

6.7. Notes and Remarks

The consideration of Banach algebras which are function algebras falls largely under the shadow of the Banach-Stone theorem and its generalizations which we have previously discussed in Chapter 2. Hence, in thinking about a chapter on Banach algebras we are naturally drawn to the influential paper of Kadison [158] in which he characterizes the linear isometries on (possibly) noncommutative C^*-algebras. Kadison's theorem, 6.1.1 and 6.2.5, plays much the same role in the noncommutative theory of C^*-algebras as did the classical theorem of Banach and Stone for the commutative case. One interpretation of the Banach-Stone theorem is that an isometry between $C(K)$-spaces determines an algebraic-∗-isomorphism between the algebras, which is the point of view in the papers of de Leeuw, Rudin, and Wermer [85] and Nagasawa [231]. In Kadison's theorem we see this same form, although it is not an algebra isomorphism that is obtained, but what Kadison called a C^*-isomorphism. We have selected two different proofs to present, one of them by Paterson [243] given in Section 2, and another by Werner [320] in Section 3. It is certainly worthwhile, however to read Kadison's original.

The wealth of generalizations of Kadison's theorem that followed will be discussed briefly in the remarks on Section 2. We do want to mention some other results here that go in somewhat different directions.

Let (Ω, μ) denote a Legesgue measure space with $\mu(\Omega) = 1$, and let $\mathcal{M}$ denote the maximal abelian self adjoint subalgebra of $\mathcal{L}(L^2(\mu))$ consisting of multiplications by L^∞ functions. Given $f \in L^\infty(\mu)$, the corresponding element in $\mathcal{M}$ is denoted by L_f^Φ. Let α be a measure preserving automorphism of the measure space and let U_α be defined by $U_\alpha f = f \circ \alpha^{-1}$. Let N_α denote the linear subspace spanned by $\mathcal{M}$ and U_α. Hopenwasser [**136**] proved the following theorem.

6.7.1. THEOREM. *Let T be a linear isometry of N_α onto N_β such that $T(I) = I$. Then there exist a measure preserving automorphism γ of the measure space and a unimodular complex number z such that*

$$T(U_\alpha) = zU_\beta \quad and,$$

$$T(L_f) = U_\gamma L_f U_\gamma^{-1}, \quad for\ all\ L_f \in \mathcal{M}.$$

Next we state an extension theorem due to Hopenwasser and Plastiras [**137**]. For a Hilbert space $\mathcal{H}$, let $\mathcal{K}(\mathcal{H})$ denote the compact operators on $\mathcal{H}$.

6.7.2. THEOREM. *Let $T : \mathcal{K}(\mathcal{H}) \to \mathcal{L}(\mathcal{H})$ be a linear isometry with the property that*

$$\sup\{|\langle T(K)x, y\rangle| : K \in \mathcal{K}\} = \|x\|\|y\|,$$

for all $x, y \in \mathcal{H}$. Then T has a unique extension to an isometry on $\mathcal{L}(\mathcal{H})$.

Davidson and O'Donovan [**239**] extended this "extension" as follows.

6.7.3. THEOREM. *Let $\mathcal{A}$ be any subspace of bounded operators on $\mathcal{H}$ which contains the compact operators. Let T be an isometry of $\mathcal{A}$ into $\mathcal{L}(\mathcal{H})$ with the property*

$$\sup\{|\langle T(A)x, y\rangle| : A \in \mathcal{A}, \|A\| = 1\} = \|x\|\|y\|,$$

for all $x, y \in \mathcal{H}$. Then T is the restriction of either a $*$-automorphism or a $*$-anti-automorphism of $\mathcal{L}(\mathcal{H})$, followed by a multiplication by a unitary.

Another class of algebras for which isometries have been determined is the class of tridiagonal algebras. An algebra $\mathcal{A}$ is *tridiagonal* if there exists a countable partition $\{E_i\}$ of the Hilbert space $\mathcal{H}$ so that every $A \in \mathcal{A}$ is block diagonal with respect to the sequence $E_1, E_2, \ldots$, i.e., for every $A \in \mathcal{A}$ it is true that $AE_i \subset E_{i-1} \oplus E_i \oplus E_{i+1}$. Jo [**154**] proved that an isometry T from a tridiagonal algebra onto itself must be of the form $T(A) = WAV$, where W, V are unitary operators. The proof is similar in spirit to one given by Moore and Trent [**224**] for the isometries of *nest algebras*.

A family $\mathcal{R}$ of projections in $\mathcal{L}(\mathcal{H})$ is called a *nest* if it is totally ordered by the relation $P < Q$ whenever $P(\mathcal{H}) \subset Q(\mathcal{H})$ and contains both 0 and I. If the nest is closed in the strong operator topology then it is called a *complete* nest. The *nest algebra*, $alg\mathcal{R}$ associated with the nest $\mathcal{R}$ is defined to be the set of all operators in $\mathcal{L}(\mathcal{H})$ which leave invariant the range space of each projection in $\mathcal{R}$. We state now a theorem first proved by Moore and Trent, but also proven by Arazy and Solel [**15**].

6.7.4. THEOREM. *Let $\mathcal{A}$ and $\mathcal{B}$ denote complete nests in $\mathcal{L}(\mathcal{H})$ and let T be a linear isometry from $alg\mathcal{A}$ onto $alg\mathcal{B}$. Then there are unitary operators U in $\mathcal{L}(\mathcal{H})$ and $V = T(I)$ in the commutant of $\mathcal{B}$ such that one of the following cases holds.*

(i) $T(L) = ULU^*V$ for every $L \in algA$ and the map $A \to UAU^* = T(A)V^*$ is an order isomorphism of $\mathcal{A}$ onto $\mathcal{B}$.

(ii) $T(L) = UJL^*JU^*V$ for every $S \in algA$ (where J is a fixed involution on $\mathcal{H}$) and the map $A \to UJAUJU^* = T(A)V^*$ is an order isomorphism on $\mathcal{A}$ onto $\mathcal{B}^\perp$.

Arazy and Solel have also characterized surjective linear isometries between unital nonselfadjoint operator algebras [15] . This gives a nonselfadjoint operator algebra version of Kadison's theorem for isometries (rather than complete isometries as discussed in Section 4).

There are, of course, many papers dealing with isometries on Banach algebras and operator algebras which we have not mentioned specifically. We list here a few more references which the interested reader might want to consult: [14], [77], [104] [146], [155], [156], [157], [225], [260], [290], and [312].

We must also mention here the book by Jarosz [147] which has a wealth of material about isometries on various Banach algebras.

Kadison's theorem. We have already discussed the significance of Kadison's theorem. As we mentioned, in his proof he characterized the extreme points of the dual ball and used the fact that the conjugate of an isometry must map extreme points to extreme points, what we have frequently called the "extreme point method." This approach has been of much interest. As an example, we refer the reader to a paper of Labuschagne and Mascioni [190] which studies operators on C^*-algebras that have the extreme point preserving property.

The proof of equation (112) can be found in Bonsall and Duncan [42, p.34] as can the proof of Theorem 6.2.1. (In particular see pages 46 and following in that book.) The proofs of Lemmas 6.2.2, 6.2.3 and part of 6.2.4 are adapted from Takesaki [299], while those of 6.2.4 and Theorem 6.2.5 come principally from Paterson [243].

The first generalization of Kadison's theorem to C^*-algebras without identity seems to be Theorem 6.2.6, which is due to Paterson and Sinclair [244]. In fact Harris [128] had such a generalization, but in his characterization, the *- isomorphism does not go back into the original algebra. (See the end of the next paragraph.)

A mountain of literature and important developments has grown up around the fact that although the map τ in the Kadison characterization of an isometry T is not an algebra homomorphism, it does preserve the involution and as well as powers of elements, the so-called *quantum mechanical structure* of the C^*-algebras [158]. Perhaps one of the earliest developments is that of Harris [128], who generalized Kadison's result to Banach subspaces of $\mathcal{L}(\mathcal{H})$ which are not necessarily *-closed. Harris defined a power algebra $\mathcal{A}$ to be a Banach subspace of $\mathcal{L}(\mathcal{H})$ which contains the identity I and satisfies the condition $a \in \mathcal{A}$ implies $a^2 \in \mathcal{A}$. He obtained the Kadison conclusion for power algebras with a combination of methods from analytic function theory and some results from Kadison's paper, although his ∗-isomorphism is between the initial power algebra and an associated power algebra to the original range.

Harris obtained a further generalization in 1973 when he defined a closed subspace of the bounded operators on one Hilbert space to another as a J^*-algebra if it was closed under the operation $A \to AA^*A$ and a bounded linear bijection between two J^*-algebras was called a J^*-isomorphism if it satisfied $J(AA^*A) = J(A)J(A)^*J(A)$. These definitions, of course, apply to closed subspaces of C^*-algebras, and sometimes the operations considered are the Jordan product and the Jordan triple product which we defined in the text. Of course, J^*-algebras are not always actual algebras, and they came later to be called JC^*-triples. Harris proved the following theorem [**129**].

6.7.5. THEOREM. *Let T be a surjective isometry of a J^*-algebra $\mathcal{A}$ onto a J^*-algebra $\mathcal{B}$. Then T is a J^*-isomorphism, i.e.,*

$$T(xy^*z + zy^*x) = Tx(Ty)^*Tz + Tz(Ty)^*Tx$$

for all $x, y, z \in \mathcal{A}$.

Harris made heavy use of holomorphic methods and showed that the open unit ball of a J^*-algebra is a bounded symmetric domain.

A complex Jordan algebra $\mathcal{A}$ is a Jordan C^*-algebra if it is a Banach space whose norm satisfies (i)$\|x \circ y\| \le \|x\|\|y\|$, (ii) $\|x\| = \|x^*\|$, and (iii)$\|xx^*x\| = \|x\|^3$ for all $x \in \mathcal{A}$. A Jordan algebra is, of course, defined to be closed under the Jordan product

$$x \circ y = \frac{1}{2}(xy + yx).$$

Jordan C^*-algebras are also called JB^*-algebras. Kaplansky had conjectured that if T is a surjective isometry between unital Jordan C^*-algebras, then T is a Jordan ∗- isomorphism, i.e.,

$$T(x \circ y) = Tx \circ Ty, \text{ and } T(x^*) = (Tx)^*.$$

Wright and Youngson [**324**] proved this conjecture.

Wright and Youngson also noted that the same result had been obtained earlier by Kaup [**169**] using much different methods.

Let us mention now a class which includes the ones previously discussed, and which arises in the study of bounded symmetric domains in complex Banach spaces [**169**] and appears also as the range of contractive projections on C^*-algebras [**109**]. A JB^*-triple is a complex Banach space X endowed with a continuous sesqui-linear map $D : X \times X \to \mathcal{L}(X)$ such that $D(x, x)$ is Hermitian positive, $\|D(x, x)\| = \|x\|^2$, and if we define $\{xyz\} = D(x, y)z$, then $\{xyz\} = \{zyx\}$ and furthermore

$$\{xy\{uvz\}\} + \}u\{yxv\}z\} = \{\{xyu\}vz\} + \{uv\{xyz\}\}.$$

Every C^*-algebra is a JB^*-triple with $\{xyz\} \equiv (1/2)(xy^*z + zy^*x)$ and a Jordan C^*-algebra is a JB^*-triple with

$$\{xyz\} = (x \circ y^*) \circ z + (z \circ y^*) \circ x - (z \circ x) \circ y^*.$$

The following theorem of Kaup [169] generalizes all the ones we have previously mentioned.

6.7.6. THEOREM. *Let T be a surjective linear isometry from a JB^*-triple $\mathcal{A}$ onto a JB^*-triple $\mathcal{B}$. Then T is a JB^*-triple isomorphism, i.e.,*

$$T\{xyz\} = \{(Tx)(Ty)(Tz)\} \ \text{ for all } \ x, y, z \in \mathcal{A}.$$

Thus, as Werner [320] puts it, two JB^*-triple systems are isometrically isomorphic as Banach spaces if and only if they are isomorphic as JB^*-triple systems. Indeed, because of the connection with bounded symmetric domains, the open unit balls of two Banach spaces X and Y are biholomorphically equivalent if and only if X and Y are isometrically isomorphic [170].

An excellent discussion of these matters is given by Dang, Friedman, and Russo [82] who have given a new proof which depends on the affine geometric properties of the convex set of states instead of pure states or extreme points.

We conclude this subsection by mentioning a few more pertinent references: [69], [76], [81], [108], [130], [131], [141], [142] [190], [306] [307],[308], and [309].

Subdifferentiability and Kadison's theorem. The material in this section is entirely the work of W. Werner [320] with a little help from his colleagues Contreres, Paya, and Taylor. It is very interesting that Werner was able to use differentiability of the norm in this much less hospitable setting, because it touches base with Banach's original proof. Lemma 6.3.1 is crucial and it reminds one of the result of Russo and Dye [278] who proved that a linear mapping of one C^*-algebra to another, which maps unitaries into unitaries can be written as a product of a unitary with a JB^* homomorphism.

Lemma 6.3.3, which plays a pivotal role in the proofs of 6.3.5, 6.3.6, and 6.3.7, is due to Gregory [117]. It is through this lemma that the subdifferentiability of the norm is brought to bear. The remarks prior to the statement of this lemma and the proof of the lemma are adaptations of arguments from the book of Phelps [252] and the paper of Gregory cited above. Lemma 6.3.4 may be found in the paper of Taylor and Werner [303], while 6.3.5 and 6.3.6 come from Contreres, Paya, and Werner [78] and the reader should consult those papers to fill in some of the details we omitted. Lemma 6.3.7 and the proof of Kadison's theorem are taken from [320].

Although it is not at all related to what is going on in this section, there exists a paper by Bachir using differentiablity to extend the classical Banach-Stone theorem to complete metric spaces [19].

The nonsurjective case of Kadison's theorem. The goal of "noncommutative functional analysis" seems to be the study of operator spaces as a generalization of Banach spaces. The names of Effros and Arveson, appear to be prominent in its early development. At this time there are two major treatises on the subject, one by Effros and Ruan [**91**] and the other by Pisier [**253**]. The book of Paulsen [**247**] is very helpful. Because of the importance of this theory in functional analysis today, we thought it worthwhile to use it as the setting for our discussion of Kadison's theorem in the nonsurjective case.

Our principal source for the material in this section comes from a preprint of a paper by Blecher and Hay [**38**]. The definitions of operator system, operator space, completely bounded, completely positive, and completely isometric were taken from Paulsen [**247**], and the example mentioned of a positive, contractive map which fails the corresponding complete properties can also be found there. The proof of Proposition 6.4.1 is adapted from the proof of Proposition 3.7 and Theorem 3.8 in [**247**]. The proof of Proposition 6.4.2 is based on an argument in Pedersen [**248**, p.16].

The reader will have noted that a triple system is similar to what Harris [**129**] called a J^*-algebra. A triple morphism, as we have defined it, is not the same thing as a JB^*-homomorphism, however. We have defined a triple morphism to be bounded, but this can be dispensed with, because a linear map satisfying (116) can be shown to be bounded. The proof of Proposition 6.4.3 is taken from Harris [**129**], who was proving that a J^*-isomorphism must be an isometry. But the condition (115) defining a J^*-isomorphism (or JB^*-isomorphism) is not sufficient to extend the condition to products of matrices.

Let us insert here an interesting result about triple morphisms. It seems to be part of the folklore of the subject.

6.7.7. LEMMA. *Let $\mathcal{A}$ and $\mathcal{B}$ be unital C^*-algebras.*

(i) *A map $T : \mathcal{A} \to \mathcal{B}$ of the form $T = u\pi(\cdot)$ for a unitary $u \in \mathcal{B}$ and an algebra-$*$-homomorphism $\pi : \mathcal{A} \to \mathcal{B}$, is a triple morphism.*

(ii) *If $T : \mathcal{A} \to \mathcal{B}$ is a triple morphism, and if $u = T(1)$ and $\pi = u^*T(\cdot)$, then π is an algebra-$*$-homomorphism. Moreover, u is a partial isometry and we can write $T = u\pi(\cdot)$.*

(iii) *If $T : \mathcal{A} \to \mathcal{B}$ is a surjective triple morphism, and if u, π are as in (ii), then u is a unitary in $\mathcal{B}$, π is a surjective algebra-$*$homomorphism from $\mathcal{A}$ onto $\mathcal{B}$, and $T = u\pi(\cdot)$.*

PROOF. (i) For this simply unravel $u\pi(x)(u\pi(y))^*u\pi(z)$.

(ii) We must have $\|T\| \leq 1$ and hence $\|u\| \leq 1$. Thus π is also a contraction, and using (116), we get

$$\pi(x)\pi(y) = u^*T(x)(T(1))^*T(y) = u^*T(xy) = \pi(xy)$$

and π is a homomorphism. Since a contractive homomorphism between C^*-algebras is necessarily a $*$-homomorphism, we have shown the first part of (ii). However, because T is a triple morphism, we must have $uu^*u = u$ so that u is a partial isometry and

$$u\pi(\cdot) = uu^*T(\cdot) = T(\cdot).$$

(iii) For $z \in \mathcal{A}$,

$$uu^*T(z) = T(1)T(1)^*T(z) = T(z),$$

and since T is surjective, we can take $T(z) = 1$ to get $uu^* = 1$. Now $u^*u = 1$ by the same argument, so u is unitary. The rest is clear. □

It is known that if T is a surjective map and is a complete isometry, then T is a triple morphism. This fact appears to have been proved independently by Hamana [125], Kirchberg [171], and Ruan [271], and first appeared in Ruan's Ph.D. thesis. Indeed, from the beginning of the proof of (ii) in Theorem 6.4.4, we get $\mu(T(a)) = a$ where μ is a triple morphism. Since T is surjective, we can apply T^{-1} to conclude that T is a triple morphism. From this and Proposition 6.4.3, we get the following nice statement.

6.7.8. PROPOSITION. *A surjective linear map between triple systems is a complete isometry if and only if it is a triple morphism.*

This, together with 6.7.7, would lead to the following version of Kadison's theorem.

6.7.9. THEOREM. *A surjective linear map $T : \mathcal{A} \to \mathcal{B}$ between unital C^* algebras is a complete isometry if and only if T is of the form $T(\cdot) = u\pi(\cdot)$ for u a unitary element of $\mathcal{B}$ and π a 1-1 $*$-homomorphism of $\mathcal{A}$ onto $\mathcal{B}$.*

We are indebted to David Blecher for discussions on these matters.

The notion of triple envelope (or what is sometimes called the *noncommutative Šilov boundary* is due to Arveson [17, 18] and Hamana [124, 125]. Another good reference here is [37], which includes an informative introduction as well as an extensive bibliography.

The main theorem, 6.4.4, is due to Blecher and Hay [38]. We have only scratched the surface of what is in their paper which includes much information about triple systems and complete isometries. Another related reference is [39]. We note that the projection p in (iii) and (iv) of the theorem can be chosen so that $1 - p$ is closed and, in fact, one could take $p = u^*u$. For a discussion of "closed projections," see [248, Sec. 3.11].

It is important to mention another paper which we have in preprint form and which treats nonsurjective isometries between C^*-algebras. This work is by Chu and Wong [70], and is not concerned with complete isometries. By the work of Kaup [169], which we discussed earlier in subsection 2, the geometry of bounded symmetric domains has been shown to be completely determined by the Jordan triple structures of the related C^*-algebras, or JB^*-triples. Remarking on this, Chu and Wong indicate the importance of the Jordan triple product in the study of isometries on C^* algebras. In particular, they ask to what extent the triple preserving property of an isometry persists if it is not surjective.

Let us state one of the main results from their paper. For the purposes of this statement, we will write

$$\{x, y, z, \} = \frac{1}{2}(xy^*z + zy^*x)$$

for x, y, z elements of a C^*-algebra.

6.7.10. THEOREM. *(Chu and Wong) Let $\mathcal{A}$ and $\mathcal{B}$ be C^*-algebras and let $T : \mathcal{A} \to \mathcal{B}$ be a linear isometry. Then there is a largest projection p in $\mathcal{B}^{**}$, called the structure projection of T, such that*

(i) $T(\cdot)p : \mathcal{A} \to \mathcal{B}^{**}$ *is a triple homomorphism;*
(ii) $T\{a, b, c\}p = \{Ta, Tb, Tc\}p$ *for all a, b, c in $\mathcal{A}$.*
Further, p is a closed projection and $(Ta)^*(Tb) = p(Ta)^*(Tb)$ for all $a, b \in \mathcal{A}$. If $\mathcal{A}$ is abelian, then $\|T(a)p\| = \|a\|$ for all $a \in \mathcal{A}$.

The reader should note that the triple homomorphism as in the theorem just above, is not the same thing as a triple morphism which appears in Proposition 6.4.3 or in Theorem 6.4.4. In fact, the difference between the two notions pinpoints the difference between an isometry and a complete isometry. Also, we should remark that in the abelian case, since $T(\cdot)p$ is an isometry, Theorem 6.7.10 extends Holsztynski's theorem in the same way as we have seen before.

The algebras $C^{(1)}$ and AC. The material in this section is taken entirely from the 1965 paper of Cambern [**50**]. As was pointed out by Jarosz [**147**], the theorems in this paper of Cambern were the first generalizations of the theorem of Nagasawa [**231**] (or de Leeuw, Rudin, and Wermer [**85**]) to some nonuniform algebras. It led to a number of other papers. Rao and Roy [**263**] proved the same result in 1971 for algebras of Lipschitz functions and continuously differentiable functions. Cambern and Pathak [**55, 56**], and Pathak [**245, 246**] obtained results on $C^{(1)}(X)$ and $AC(X)$ where X is a certain subset of $[0, 1]$. Other related papers include [**120**], [**213**], [**214**], [**258**], [**264**], [**315**], [**316**], and [**318**].

In 1988, Jarosz and Pathak [**150**] devised a scheme by which it could be verified whether an isometry between certain subspaces of $C(Q)$-spaces is given by a homeomorphism between the corresponding compact Hausdorff spaces. This scheme could be used to obtain the results we have previously mentioned as well as many others. Here is what they set up.

Let X be a subspace of a Banach space $C(Q)$, which separates the points of Q, and let T_X be a linear map from X into a Banach space Y. It is assumed that the complete norm on X is given by one of the following formulas: for $f \in X$,

(M) $\|f\| = max(\|f\|_\infty, \|T_X f\|)$;

(Σ) $\|f\| = \|f\|_\infty + \|T_X f\|$;

(C) $\|f\| = \sup\{|f(t)| + |T_X f(t)| : t \in Q\}$,

where in the last case, it is assumed that $Y = C(Q)$. A subspace X is said to be an M-subspace of Q, Σ-subspace of Q, or a C-subspace of Q, if the norm on X is given by the formula (M), (Σ), or (C), respectively, plus a number of other conditions too numerous to list here. Here is one of the theorems.

6.7.11. THEOREM. *Let X and Y be (M)-subspaces of $C(Q)$ and $C(K)$, respectively. Let $Q_0 = \{\alpha\psi_s : s \in ch(X), |\alpha| = 1|\}$, $K_0 = \{\alpha\psi_t : t \in ch(Y), |\alpha| = 1\}$. Then an isometry T from X onto Y is canonical if and only if $T^*(K_0) = Q_0$.*

It is pretty clear from what we have previously seen, why this works. Jarosz and Pathak point out that for the classical function spaces with M-norm, the assumption can be easily verified.

Douglas algebras. Douglas algebras were named after R. Douglas, who conjectured that any closed subalgebra $\mathcal{A}$ of L^∞ that contains H^∞ is generated by H^∞ and a set of inverses of inner functions [73]. The conjecture was later proved by Marshall and Chang [211, 64]. Perhaps the most famous proper Douglas algebra is the algebra $H^\infty + C$.

The proofs are taken from the paper of Font [105], who is actually concerned with the more general problem of describing the isometries between linear subspaces of L^∞ which contain H^∞. The important Theorem 2.5.3, which is stated in the notes of Chapter 2, is due to Araujo and Font [10]. It enables us to consider Šilov boundaries, about which more is known, instead of the Choquet boundaries which are used in the theorems of Chapter 2 to get the nice weighted composition form for the isometries. A key fact here is that two Douglas algebras have the same Šilov boundary. This is true because $\partial H^\infty = \partial L^\infty$. A proof of this latter statement can be found in [112, p. 191] or in [132, p. 172].

It is well known that any algebra isomorphism on a $C(K)$-space is given by a composition $f \to f \circ \varphi$, and so using the Gelfand theory, and the fact that $L^\infty = C(K)$, where K is the maximal ideal space of L^∞, it is not hard to see that an algebra isomorphism of L^∞ is as described in Theorem 6.6.1. We thought the direct proof given by Font was interesting.

In the paper cited [105], Font also proves that any linear bijection on L^∞ which preserves nonvanishing functions (i.e., functions f such that $f(t) \neq 0$ for every $t \in \mathbf{T}$), must be of the form

$$Tf = h(f \circ Sj)$$

where $h \in L^\infty$ and S is an algebra isomorphism of L^∞ onto itself induced by T.

Bibliography

[1] Y. Abramovich, *Multiplicative representation of disjointness preserving operators*, Nederl. Akad. Wetensch. Indago. Math. **48** (1983), 265–279.

[2] Y. Abramovich and M. Zaidenberg, *A rearrangement invariant space isometric to L_p concides with L_p*, Institute Fourier (Grenoble) **287** (1994), 1–6.

[3] P. Ahern and R. Schneider, *Isometries of H^∞*, Duke Math. J. **42** (1975), 321–326.

[4] A. Al-Hussaini, *Potential operators and equimeasurability*, Pac. J. Math. **76** (1978), 1–7.

[5] D. Amir, *On isomorphisms of continuous function spaces*, Israel J. Math. **3** (1965), 89–173.

[6] J. Anderson, J. Clunie, and C. Pommerenke, *On Bloch functions and normal functions*, J. Reine Angew. Math. **270** (1974), 12–37.

[7] K.F. Anderson, *On L^p norms and the equimeasurability of functions*, Proc. Amer. Math. Soc. **40** (1973), 149–153.

[8] T. Anderson, *Statistical analysis of time series*, Wiley, New York, 1971.

[9] T. Ando, *Contractive projections in L^p spaces*, Pac. J. Math. **17** (1966), 391–405.

[10] J. Araujo and J. Font, *Linear isometries between subspaces of continuous functions*, Trans. Amer. Math. Soc. **349** (1997), 413–428.

[11] J. Araujo and J. Font, *On Shilov boundaries for subspaces of continuous functions*, Topology Appl. **77** (1997), 79–85.

[12] J. Araujo and K. Jarosz, *Isometries of spaces of unbounded continuous functions*, Bull. Austral. Math. Soc. **63** (2001), 475–484.

[13] J. Arazy, *Isometries of complex symmetric sequence spaces*, Math. Zeit. **188** (1985), 427–431.

[14] J. Arazy, *Isometries of Banach algebras satisfying the von Neumann inequality*, Math. Scand. **74** (1994), 137–151.

[15] J. Arazy and B. Solel, *Isometries of non-self-adjoint operator algebras*, J. Funct. Anal. **90** (1990), 284–305.

[16] R. Arens and J. Kelley, *Characterization of spaces of continuous functions over compact Hausdorff space*, Trans. Amer. Math. Soc. **62** (1947), 499–508.

[17] W. Arveson, *Subalgebras of C^*-algebras*, Acta Math. **123** (1969), 141–224.

[18] W. Arveson, *Subalgebras of C^*-algebras II*, Acta Math. **128** (1972), 271–308.

[19] M. Bachir, *Sur la différentiabilité générique et le théorème de Banach-Stone*, C.R. Acad. Sci. Paris **330** (2000), 687–690.

[20] J. Baker, *Isometries in normed spaces*, American Math. Monthly **78** (1971), 655–658.

[21] S. Banach, *Theorie des operations lineares*, Chelsea, Warsaw, 1932.

[22] H. Bauer, *Un probleme de Dirichlet pour la frontiere de Šilov d'un espace compact*, C. R. Acad. Sci., Paris **247** (1958), 843–846.

[23] E. Behrends, *M structure and the Banach-Stone theorem*, Lecture Notes in Mathematics, vol. 736, Springer-Verlag, Berlin and New York, 1979.

[24] C. Bennett and R. Sharpley, *Interpolation of operators*, Academic Press, Boston, 1988.

[25] Y. Benyamini, *Small into isomorphisms between spaces of continuous functions*, Proc. Amer. Math. Soc. **83** (1981), 479–485.

[26] E. Berkson, *One-parameter semigroups of isometries into H^p*, Pacific J. Math. **86** (1980), 403–413.

[27] E. Berkson, *On spectral families of projections in Hardy spaces*, J. Funct. Anal. **60** (1985), 146–167.

[28] E. Berkson, R. Kaufman, and H. Porta, *Mobius transformations of the disc and one-parameter groups of isometries of H^p*, Trans. Amer. Math. Soc. **199** (1974), 223–239.

[29] E. Berkson and H. Porta, *Hermitian operators and one-parameter groups of isometries in Hardy spaces*, Trans. Amer. Math. Soc. **185** (1973), 331–334.

[30] E. Berkson and H. Porta, *One-parameter groups of isometries on Hardy spaces of the torus*, Trans. Amer. Math. Soc. **220** (1976), 373–391.

[31] E. Berkson and H. Porta, *One-parameter groups of isometries on Hardy spaces of the torus: spectral theory*, Trans. Amer. Math. Soc. **227** (1977), 357–370.

[32] E. Berkson and H. Porta, *The group of isometries of H_p*, Ann. Mat. Pura Appl. **119** (1979), 231–238.

[33] E. Berkson and H. Porta, *The group of isometries on Hardy spaces of the n-ball and the polydisc*, Glasgow Math. J. **21** (1980), 199–204.

[34] R. Billingsley, *Probability and measure*, third ed., Wiley-Interscience, New York, 1995.

[35] E. Bishop, *A minimal boundary for function algebras*, Pac. J. Math. **9** (1959), 629–642.

[36] E. Bishop and K. deLeeuw, *The representation of linear functionals by measures on sets of extyreme points*, Ann. Innst. Fourier, Garenoble **9** (1959), 305–331.

[37] D. Blecher, *The Shilov boundary of an operator space and the characterization theorems,*, J. Funct. Anal. **182** (2001), 280–343.

[38] D. Blecher and D. Hay, *Complete isometries into C^*-algebras*, preprint.

[39] D. Blecher and L. Labuschagne, *Logmodularity and isometries of operator algebras*, preprint.

[40] R. Boas, *Isomorphism between H_p and L_p*, Amer.J.Math **77** (1955), 655–656.

[41] S. Bochner, *Inversion formulae and unitary transformations*, Annals of Math. **35** (1934), 111–115.

[42] F. Bonsall and J. Duncan, *Numerical ranges of operators on normed spaces and of elements of normed algebras*, London Mathematical Society Lecture Note Series, Cambridge University Press, London, 1971.

[43] A. Bosznay, *On a theorem of Mazur and Ulam*, Periodica Math. Hungarica **16** (1985), 7–13.

[44] J. Bourgain, *Real isomorphic complex Banach spaces need not be complex isomorphic*, Proc.Amer.Math.Soc. **96** (1986), 221–226.

[45] D. Bourgin, *Approximately isometric and multiiplicative transformations on continuous function rings*, Duke Math. J. **16** (1949), 385–397.

[46] D. Bourgin, *Classes of transformations and bordering transformations*, Bull. Amer. Math. Soc. **57** (1951), 223–237.

[47] B. Brosowski and F. Deutsch, *On some geometric properties of suns*, J. Approx. Theory **10** (1974), 245–267.

[48] R. Buck, *A complete characterization for extreme functionals*, Bull. Amer. Math. Soc. **65** (1959), 130–133.

[49] R. Buck, *Applications of duality in approximation theory*, Approximation of functions (J. Garabedian, ed.), Elsevier, Amsterdam, 1965.

[50] M. Cambern, *Isometries of certain Banach algebras*, Studia Math. **25** (1965), 217–225.

[51] M. Cambern, *A generalized Banach-Stone theorem*, Proc. Amer. Math. Soc. **17** (1966), 396–400.

[52] M. Cambern, *On isomorphisms with small bound*, Proc. Amer. Math. Soc. **18** (1967), 1062–1066.

[53] M. Cambern, *On mappings of madules, with application to spaces of analytic functions*, Bull. Inst. Math. Acad. Sinica **5** (1977), 105–119.

[54] M. Cambern, *A Holsztynski theorem for spaces of vector valued functions*, Studia Math. **63** (1978), 213–217.

[55] M. Cambern and V. Pathak, *Isometries of spaces of differentiable functions*, Math. Japon. **26** (1981), 253–260.

[56] M. Cambern and V. Pathak, *Isometries of spaces of differentiable functions*, Rev. Roumaine Math. Pures Appl. **22** (1982), 737–743.

[57] S. Campbell, G. Faulkner, and M. Gardner, *Isometries on L^p spaces and copies of l^p shifts*, Proc. Amer. Math. Soc. **77** (1979), 178–200.

[58] S. Campbell, G. Faulkner, and R. Sine, *Isometries, projections and Wold decompositions*, Pitman Res. Notes in Math. **38** (1979), 85–114.

[59] N. Carothers, S. Dilworth, and D. Troutman, *On the geometry of the unit sphere of the Lorentz space $L_{w,1}$*, Glasgow Math. J. **34** (1992), 21–25.

[60] N. Carothers, R. Hayden, and P. Lin, *On the isometries of the Lorentz spaces $L_{w,p}$*, Israel J. Math. **84** (1993), 265–287.

[61] N. Carothers and B. Turrett, *Isometries on $L_{p,1}$*, Trans. Amer. Math. Soc. **297** (1986), 95–103.

[62] P. Cembranos and J. Mendoza, *Banach spaces of vector-valued functions*, Lecture Notes in Mathematics, vol. 1676, Springer-Verlag, Berlin, 1997.

[63] B. Cengiz, *On extremely regular function spaces*, Pac. J. Math. **49** (1973), 335–338.

[64] S. Chang, *A characterization of Douglas subalgebras*, Acta Math. **137** (1976), 82–89.

[65] Z. Charzynski, *Sur les transformations isometriques des espaces du type (F)*, Studia Math. **13** (1953), 217–225.

[66] M. Chasles, Bull. des Sciences Mathematiques de Ferrussae **XXIV** (1831), 321.

[67] J. Choksi, *Unitary operators induced by measure preserving transformations*, J. of Math. and Mech. **16** (1966), 83–100.

[68] J. Choksi, *Unitary operators induced by measurable transformations*, J. of Math. and Mech. **17** (1968), 785–801.

[69] C. Chu, T. Dang, B. Russo, and B. Ventura, *Surjective isometries of real C^*-algebras*, J. London Math. Soc. **47** (1993), 97–118.

[70] C. Chu and N. Wong, *Isometries between C^*-algebras*, Revista Matemctica Iberoamericana, to appear.

[71] J. Cima and W. Wogen, *Extreme points of the unit ball of the Bloch space B_0*, Mich. Math. J. **25** (1978), 213–222.

[72] J. Cima and W. Wogen, *On isometries of the Bloch spaces*, Illinois J. Math. **24** (1980), 313–316.

[73] K. Clancey and W. Cutrer, *Subalgebras of Douglas algebras*, Proc. Amer. Math. Soc. **40** (1973), 102–106.

[74] J. Clarkson, *Uniformly convex spaces*, Trans. Amer. Math. Soc. **46** (1936), 396–411.

[75] J. Clarkson, *A characterization of C-spaces*, Ann. of Math. **48** (1947), 845–850.

[76] B. Cole and J. Wermer, *Isometries of certain operator algebras*, Proc. Amer. Math. Soc. **124** (1996), 3047–3053.

[77] J. Connor and I. Loomis, *Isometries on conservative subalgebras of bounded sequences*, Proc. Amer. Math. Soc. **107** (1989), 743–749.

[78] M. Contreras, R. Payá, and W. Werner, *C^*-algebras that are [i]-rings*, J. Math. Anal. Appl. **198** (1996), 227–236.

[79] J. Coolidge, *A history of geometrical methods*, Oxford University Press, Oxford, 1940.

[80] N. Cutland and G. Zimmer, *A new proof of the Banach-Stone theorem*, Bull. Austral. Math. Soc. **57** (1998), 55–58.

[81] T. Dang, *Real isometries between JB^* triples*, Proc. Amer. Math. Soc. **114** (1992), 971–980.

[82] T. Dang, Y. Friedman, and B. Russo, *Affine geometric proofs of the Banach-Stone theorem of Kadison and Kaup*, Rocky Mountain J. Math. **20** (1990), 409–428.

[83] M. Day, *Normed linear spaces*, second ed., Springer-Verlag, Berlin-Heidelberg-New York, 1970.

[84] K. deLeeuw and W. Rudin, *Extreme points and extremum problems in H_1*, Pacific J. Math. **8** (1958), 467–485.

[85] K. deLeeuw, W. Rudin, and J. Wermer, *The isometries of some function spaces*, Proc. Amer. Math. Soc. **11** (1960), 694–698.

[86] J. Doob, *The elementary Gaussian processes*, Ann. Math. Stat. **15** (1944), 229–282.

[87] J. Doob, *Stochastic processes*, Wiley, New York, 1953.

[88] L. Drewnowski, *On nonlinear isometries between Banach spaces*, Functional Analysis: Proceedings of the First International Workhip, Trier, 1994 (S. Dierolf, S. Dineen, and P. Domanski, eds.), de Gruyter, Berlin, 1996, pp. 129–141.

[89] N. Dunford and J. Schwartz, *Linear operators, Part I*, Interscience Publishers, New York, 1958.

[90] M. Edelstein, *On non-expansive mappings of Banach spaces*, Proc. Camb. Phil. Soc. **60** (1964), 439–447.

[91] E. Effros and Z. Ruan, *Operator spaces*, Oxford University Press, Oxford, 2000.

[92] S. Eilenberg, *Banach space methods in topology*, Annals of Math. **43** (1942), 568–579.

[93] M. El-Gebeily and J. Wolfe, *Isometries of the disc algebra*, Proc. Amer. Math. Soc. **93** (1985), 697–702.

[94] A. Ellis, *Isometries of self-adjoint complex function spaces*, Math. Proc. Cambridge Philos. Soc. **105** (1989), 133–138.

[95] A. Ellis and W. So, *Isometries and the complex state spaces of uniform algebras*, Math. Z. **195** (1987), 119–125.

[96] L. Euler, *Formulae generales pro larslaticue quacunque corporum rigidorum*, Novi Commentarii Academiae Petrysolitarrae **XX** (1776), 137–273.

[97] K. Fan, *Partially ordered additive groups of continuous functions*, Ann. of Math. **51** (1950), 409–427.

[98] G. Faulkner and J. Huneycutt, *Orthogonal decomposition of isometries in a Banach space*, Proc. Amer. Math. Soc. **69** (1978), 125–128.

[99] T. Figiel, *On non-linear isometric embeddings of normed linear spaces*, Bull. Acad. Polon. Sci. **16** (1968), 185–188.

[100] S. Fisher, *Function theory on planar domains*, John Wiley & Son, New York, 1983.

[101] R. Fleming and J. Jamison, *Isometries on certain Banach spaces*, J. London Math. Soc. (2) **9** (1974), 363–371.

[102] R. Fleming and J. Jamison, *Isometries on Banach spaces: a survey*, Analysis, geometry, and groups, a Riemann legacy Volume (H. Srivastava and T. Rassias, eds.), Hadronic Press, Palm Harbor, Florida, 1993, pp. 52–123.

[103] R. Fleming, J. Jamison, and A. Kaminska, *Isometries of Musielak-Orlicz spaces*, Proceedings of the Conference on Function Spaces (Edwardsville, Illinois) (K. Jarosz, ed.), 1992, pp. 139–154.

[104] J. Font, *Isometries between function algebras with finite codimensional range*, Manuscripta Math. **100** (1999), 13–21.

[105] J. Font, *On weighted composition operators between spaces of measurable functions*, Quaestiones Mathemaaticae **22** (1999), 143–148.

[106] F. Forelli, *The isometries of H^p*, Canad. J. Math. **16** (1964), 721–728.

[107] F. Forelli, *A theorem on isometries and the application of it to the isometries of $H^p(S)$ for $2 < p < \infty$*, Canad. J. Math. **25** (1973), 284–289.

[108] T. Franzoni, *The group of homeomorphic automorphisms in certain J^*-algebras*, Ann. Mat. Pura Appl. (4) **127** (1982), 52–66.

[109] Y. Friedman and B. Russo, *Solution of the contraction projection problem*, J. Funct. Anal. **60** (1985), 56–79.

[110] T. Gamelin, *Uniform algebras*, Prentice Hall, Englewood Cliffs, New Jersey, 1969.

[111] D. Garling and P. Wojtaszczyk, *Some Bargmann spaces of analytic functions*, Proceedings of the Conference on Function Spaces, Edwardsville, Lecture Notes in Pure and Applied Mathematics, vol. 172, Marcel Dekker, 1995.

[112] J. Garnett, *Bounded analytic functions*, Academic Press, 1981.

[113] K. Geba and Z. Semadeni, *Spaces of continuous functions (V)*, Studia Math. **19** (1960), 303–320.

[114] I. Gelfand and A. Kolmogoroff, *On rings of continuous functions on a topological space*, D. R, (Doklady) URSS **22** (1939), 11–15.

[115] J. Giles, *Classes of semi-inner product spaces*, Trans. Amer. Math. Soc. **321** (1967), 436–446.

[116] K. Goodrich and K. Gustafson, *On a converse to Koopman's lemma*, Physics **102A** (1980), 379–388.

[117] D. Gregory, *Upper semi-continuity of subdifferential mappings*, Canad. Math. Bull. **23** (1980), 11–19.

[118] R. Grzaslewicz, *Isometries of $L^1 \cap L^p$*, Proc. Amer. Math. Soc. **93** (1985), 493–496.

[119] R. Grzaslewicz and H. Schafer, *Surjective isometries of $L^1 \cap L^\infty[0,\infty]$ and $L^1 + L^\infty[0,\infty]$*, Indag. Mathem., No. 5. **3** (1992), 173–178.

[120] M. Grzesiak, *Isometries of a space of continuous functions determined by an involution*, Math. Nachr. **145** (1990), 217–221.

[121] A. Gutek, D. Hart, J. Jamison, and M. Rajagopalan, *Shift operators on Banach spaces*, J. of Functional Analysis **101** (1991), no. 1, 97–119.

[122] P. Halmos, *A Hilbert space problem book*, Van Nostrand, Princeton, 1967.

[123] P. Halmos and J. von Neumann, *Operator methods in classical mechanics II*, Ann. of Math. **43** (1942), 574–576.

[124] M. Hamana, *Injective envelopes of operator systems*, Publ. R.I.M.S. Kyoto Univ. **15** (1979), 773–785.

[125] M. Hamana, *Triple envelopes and Šilov boundaries of operator spaces*, Math. J. Toyama University **22** (1999), 77–93.

[126] C. Hardin, *Isometries on subspaces of L^p*, Indiana Univ. Math. J. **30** (1981), 449–465.

[127] G. Hardy, J. Littlewood, and G. Polya, *Inequalities*, second ed., Cambridge University Press, Cambridge, 1952.

[128] L. Harris, *Schwartz's lemma in normed linear spaces*, Proc. Nat. Acad. Sci. U.S.A. **62** (1969), 521–522.

[129] L. Harris, *Bounded symmetric homogeneous domains in infinite dimensional spaces*, Proceedings on infinite holomorphy, Lecture Notes in Math., vol. 364, Springer-Verlag, 1974, pp. 13–40.

[130] J. Herves, *On linear isometries of Cartan factors in infinite dimensions*, Ann. Mat. Pura. Appl. (4) **142** (1985), 371–379.

[131] J. Herves and J. Isidro, *Isometries and automorphisms of the spaces of spinors*, Rev. Mat. Univ. Comput. Madrid **5** (1992), 193–200.

[132] K. Hoffman, *Banach spaces of analytic functions*, Prentice Hall, Englewood Cliffs, NJ, 1962.

[133] W. Holsztynski, *Continuous mappings induced by isometries of spaces of continuous functions*, Studia Math. **26** (1966), 133–136.

[134] W. Holsztynski, *Linearization of isometric embeddings of Banach spaces, metric envelopes*, Bull. Acad. Polon. Sci. **16** (1968), 189–193.

[135] W. Holsztynski, *Lattices with real numbers as additive operators*, Dissert. Math. **62** (1969), 21–25.

[136] A. Hopenwasser, *Ergodic automorphisms and linear spaces of operators*, Duke Math.J. **41** (1974), 747–457.

[137] A. Hopenwasser and J. Plastiras, *Isometries of quasitriangular algebras*, Proc. Amer. Math. Soc. **65** (1977), 242–244.

[138] W. Hornor and J. Jamison, *Properties of isometry-inducing maps of the unit disc*, Complex Variables Theory Appl. **38** (1999), 69–84.

[139] W. Hornor and J. Jamison, *Isometries of some Banach spaces of analytic functions*, Integral Equations and Operator Theory **41** (2001), 410–425.

[140] D. Hyers and S. Ulam, *On approximate isometries on the space of continuous functions*, Ann. of Math. (2) **48** (1947), 285–289.

[141] J. Isidro, *The manifold of minimal partial isometries in the space $L(H, K)$ of bounded linear operators*, Acta Sci. Math. (Szeged) **66** (2000), 793–808.

[142] J. Isidro and A. Rodriguez-Palacios, *Isometries of JB-algebras*, Manuscripta Math. **86** (1995), 337–348.

[143] R. James, *Orthogonality and linear functionals in normed linear spaces*, Trans. Amer. Math. Soc. **61** (1947), 265–292.

[144] J. Jamison, A. Kaminska, and P. Lin, *Isometries of Musielak-Orlicz spaces II*, Studia Math. **104** (1993), 75–89.

[145] K. Jarosz, *Into isomorphisms of spaces of continuous functions*, Proc. Amer. Math. Soc. **90** (1984), 373–377.

[146] K. Jarosz, *Isometries in semisimple commutative Banach algebras*, Proc. Amer. Math. Soc. **94** (1985), 65–71.

[147] K. Jarosz, *Perturbations of Banach algebras*, Lecture Notes in Mathematics, vol. 1120, Springer-Verlag, 1985.

[148] K. Jarosz, *Automatic continuity of separating linear isomorphisms*, Canad. Math. Bull. **33** (1990), 139–144.

[149] K. Jarosz, *Isometries of Bloch spaces*, Banach spaces (Mérida, 1992), Amer. Math. Soc., Providence, RI, 1993, pp. 141–147.

[150] K. Jarosz and V. Pathak, *Isometries between function spaces*, Trans. Amer. Math. Soc. **305** (1988), 193–206.

[151] K. Jarosz and V. Pathak, *Isometries and small bound isomorphisms of function spaces*, Lecture Notes in Pure and Applied Math., vol. 136, Marcel Dekker, 1992.

[152] J. Jeang and N. Wong, *Weighted composition operators of $C_0(X)$'s*, J. of Math. Anal. and Appl. **201** (1996), 981–993.

[153] M. Jerison, *The space of bounded maps into a Banach space*, Ann. of Math. **52** (1950), 309–327.

[154] Y. Jo, *Isometries of tridiagonal algebras*, Pac. J. Math. **140** (1989), 97–115.

[155] Y. Jo and T. Choi, *Isometries of $Alg\mathcal{L}_{2n}$ and $Alg\mathcal{L}_{2n+1}$*, Kyungpook Math. J. **29** (1989), 26–36.

[156] Y. Jo and D. Ha, *Isometries of a generalized tridiagonal algebra $A_{2n}^{(m)}$*, Tsukuba J. Math. **18** (1994), 165–174.

[157] Y. Jo and I. Jung, *Isometries of A_{2n}^n*, Math. J. Toyama Univ. **13** (1990), 139–149.

[158] R. Kadison, *Isometries of operator algebras*, Ann. of Math. **54** (1951), 325–338.

[159] S. Kakutani, *Concrete representation of abstract (M)-spaces*, Ann. of Math. **42** (1941), 994–1024.

[160] N. Kalton, *An elementary example of a Banach space not isomorphic to its complex conjugate*, Canad. Math. Bull. **38** (1995), 218–222.

[161] N. Kalton and B. Randrianantoanina, *Isometries on rearrangement-invariant spaces*, C.R. Acad. Sci. Paris **316** (1993), 351–355.

[162] N. Kalton and B. Randrianantoanina, *Surjective isometries on rearrangement-invariant spaces*, Quart. J. Math. (Oxford)(2) **45** (1994), 301–327.

[163] A. Kaminska, *Isometries of Orlicz spaces*, Proceedings of Orlicz Memorial Conference (University of Mississippi), 1991.

[164] A. Kaminska, *Isometries of Musielak-Orlicz spaces equipped with the Orlicz norm*, Rocky Mountain J. Math. **24** (1994), 1475–1486.

[165] A. Kaminska, *Remarks on function spaces*, Preprint, 1997.

[166] I. Kaplansky, *Lattices of continuous functions*, Bull. Amer. Math. Soc. **53** (1947), 618–623.

[167] I. Kaplansky, *Lattices of continuous functions II*, Amer. J. Math. **70** (1948), 626–634.

[168] M. Karamata, *Sur un mode de croissance reguliere*, Bull.Math. Soc. France **61** (1933), 55–62.

[169] W. Kaup, *A Riemann mapping theorem for bounded symmetric domains in complex Banach spaces*, Math.Z. **83** (1098), 503–529.

[170] W. Kaup and H. Upmaier, *Banach spaces with biholomorphically equivlent unit balls are isomorphic*, Proc. Ameri. Math. Soc. **58** (1978), 129–133.

[171] E. Kirchberg, *On restricted perturbations in inverse images and a description of normalizer algebras in C^*-algebras*, J. Funct. Anal. **129** (1995), 1–34.

[172] D. Koehler and P. Rosenthal, *On isometries of normed linear spaces*, Studia Math. **34** (1970), 213–216.

[173] C. Kolaski, *Isometries of Bergman spaces over bounded Runge domains*, Canad. J. Math. **33** (1981), 1157–1164.

[174] C. Kolaski, *Isometries of weighted Bergman spaces*, Canad. J. Math. **34** (1982), 693–710.

[175] C. Kolaski, *Isometries of some smooth normed spaces of analytic functions*, Complex Variable Th. and Appl. **10** (1988), 115–122.

[176] C. Kolaski, *Surjective isometries of weighted Bergman spaces*, Proc. Amer. Math. Soc. **105** (1989), 652–657.

[177] A. Koldobsky, *Isometries of $L_p(X, L_q)$ and equimeasurability*, Indiana Univ. Math. J. **40** (1991), 677–705.

[178] A. Koldobsky, *Generalized Levy representation of norms and isometric embeddings into L^p-spaces*, Ann. Inst. H. Poincare (Prob. and Stat.) **28** (1992), 335–353.

[179] A. Koldobsky, *Operators preserving orthogonality are isometries*, Proc. Royal. Soc. Edinburgh **123** (1993), 835–837.

[180] A. Koldobsky, *Isometries of L_p spaces of solutions of homogeneous partial differential equations*, Proceedings of the Conference on Function Spaces, Edwardsville, 1994, Lecture Notes in Pure and Applied Mathematics, vol. 172, Marcel Dekker, 1995, pp. 251–163.

[181] A. Koldobsky and H. Konig, *Aspects of the isometric theory of Banach spaces*, Handbook of the geometry of Banach spaces (W. Johnson and J. Lindenstrauss, eds.), vol. 1, Elsevier Science B.V., 2001, pp. 899–939.

[182] A.N. Kolmogorov, *Stationary sequences in Hilbert spaces*, Bull. Moskov. Gus. Univ. Mat. **2** (1941), 1–40 (Russian).

[183] A. Koranyi and S. Vagi, *Isometries of H_p spaces of bounded symmetric domains*, Canad. J. Math. **28** (1976), 334–340.

[184] S. Krantz and D. Ma, *On isometrivc isomorphoisms of the Bloch space in the unit ball*, Michigan Math. J. **36** (1989), 73–180.

[185] M. Krasnoselskii and Y. Ruticki, *Convex functions and Orlicz spaces*, P. Noordhoff Ltd, Groningen (The Netherlands), 1961, Translation.

[186] M. Krein and S. Krein, *On an inner characteristic of the set of all continuous functions defined on a bicompact Hausdorff space*, C. R. (Doklady) Acad. Sci. URSS **27** (1940), 427–430.

[187] M. Krein and D. Milman, *On the extreme points of regularly convex sets*, Studia Math. **9** (1940), 133–138.

[188] J. Krivine, *Plongment des espaces normés dans les l^p pour $p > 2$*, C.R.Acad.Sci. Paris **261** (1965), 4307–4310.

[189] S. Kulkarni and B. Limaye, *Real function algebras*, Monographs and Textbooks in Pure and Applied Mathematics, vol. 168, Marcel Dekker, Inc., New York, 1992.

[190] L. Labuschagne and V. Mascioni, *Linear maps between C^*-algebras whose adjoints preserve extreme points of the dual ball*, Adv. Math. **138** (1998), 15–45.

[191] H. Lacey, *The isometric theory of classical Banach spaces*, Springer-Verlag, Berlin, 1974.

[192] N. Lal and S. Merrill, *Isometries of H^p spaces of the torus*, Proc. Amer. Math. Soc. **31** (1972), 465–471.

[193] J. Lamperti, *On the isometries of some function spaces*, Pacific J. Math. **8** (1958), 459–466.

[194] K. Lau, *A representation theorem for isometries of $C(X, E)$*, Pacific J. Math. **60** (1975), 229–233.

[195] A. Lazar, *Affine functions on simplexes and extreme operators*, Israel J. Math. **5** (1967), 31–43.

[196] F. Lessard, *Invertible linear operators induced by invertible point transformations*, Ann. Sci. Math. Québec **18** (1994), 199–218.

[197] P. Levy, *Theorie de l'addition de variable aléatoires*, Gauthier-Villars, Paris, 1937.

[198] P. Lin, *Maximality of rearrangement invariant spaces*, Bull. Polish Acad. Sci. **44** (1996), 381–390.

[199] J. Lindenstrauss and L. Tzafriri, *Classical Banach spaces*, Lecture Notes in Mathematics, Springer-Verlag, Berlin, 1973.

[200] J. Lindenstrauss and L. Tzafriri, *Classical Banach spaces II*, Springer-Verlag, Berlin, 1979.

[201] I. Loomis, *Isometries on Banach spaces*, Ph.D. thesis, Memphis State University, 1982.

[202] G. Lovblom, *Isometries and almost isometries between spaces of continuous functions*, Israel J. Math. **56** (1986), 143–159.

[203] G. Lumer, *Semi-inner product spaces*, Trans. Amer. Math. Soc. **100** (1961), 29–43.

[204] G. Lumer, *Isometries of reflexive Orlicz spaces*, Bull. Amer. Math. Soc. **68** (1962), 28–30.

[205] G. Lumer, *On the isometries of reflexive Orlicz spaces*, Ann. Inst. Fourier **68** (1963), 99–109.

[206] W. Lusky, *Some consequences of Rudin's paper 'L^p-isometries and equimeasurability'*, Indiana Univ. Math. J. **27** (1978), 859–866.

[207] W. Luxemburg, *Banach function spaces*, Ph.D. thesis, Delft Institute of Technology, Assen(Netherlands), 1955.

[208] W. Luxemburg, *Rearrangement-invariant Banach function spaces*, Proc. Sympos. in Analysis, Queen's Papers in Pure and Appl. Math. **10** (1967), 83–144.

[209] P. Mankiewicz, *On isometries in linear metric spaces*, Studia Math. **55** (1976), 163–173.

[210] P. Mankiewicz, *Fat equicontinuous groups of homeomorphisms of linear topological spaces and their application to the problem of isometries in linear metric spaces*, Studia Math. **64** (1979), 13–23.

[211] D. Marshall, *Subalgebras of L^∞ containing H^∞*, Acta Math. **137** (1976), 91–98.

[212] P. Masani and N. Wiener, *The prediction theory of multivariate stochastic processes, part I*, Acta Math. **98** (1957), 111–150.

[213] T. Matsumoto and S. Watanabe, *Extreme points and linear isometries of the domain of a closed * derivation of $C(K)$*, J. Math. Soc. Japan **48** (1996), 229–254.

[214] T. Matsumoto and S. Watanabe, *Surjective linear isometries of the domain of a *-derivation equipped with the Cambern norm*, Math. Z. **230** (1999), 185–200.

[215] J. Mayer, *Isometries in Banach spaces of functions holomorphic on the disc and smooth up to its boundary*, Vestnik Moskov. Univ. Ser. I Mat. Mekh. **34** (1979), 237–241.

[216] J. Mayer, *Isometries in spaces of analytic functions with bounded derivative*, Soviet Math. Dokl. **21** (1980), 135–137.

[217] J. Mayer, *Isometries in spaces of analytic functions with slowly growing derivatives*, Soviet Math. Doklady **26** (1982), 424–427.

[218] S. Mazur, *Canonical isometry on weighted Bergman spaces*, Pacific J. Math. **136** (1989), 303–310.

[219] S. Mazur and S. Ulam, *Sur les transformation d'espaces vectoriels normé*, C.R. Acad. Sci. Paris **194** (1932), 946–948.

[220] J. McDonald, *Isometries of function algebras*, Illinois J. Math. **17** (1973), 579–583.

[221] J. McDonald, *Isometries of the disk algebra*, Pacific J. Math. **58** (1975), 143–154.

[222] J. Merlo, *On isometries in L^p spaces*, Advances in Mathematics **15** (1975), 194–197.

[223] A. Milgram, *Multiplicative semigroups of continuous functions*, Duke Math. J. **16** (1949), 377–383.

[224] R. Moore and T. Trent, *Isometries of nest algebras*, J. Funct. Anal. **86** (1989), 180–210.

[225] R. Moore and T. Trent, *Isometries of certain reflexive operator algebras*, J. Funct. Anal. **98** (1991), 437–471.

[226] P. Muhly, *Isometries of ergodic Hardy spaces*, Israel J. Math. **36** (1980), 50–74.

[227] J. Musielak, *Orlicz spaces and modular spaces*, Lecture Notes in Mathematics, vol. 1034, Springer-Verlag, Berlin, 1983.

[228] S. Myers, *Banach spaces of continuous functions*, Ann. of Math. **49** (1948), 132–140.

[229] S. Myers, *Spaces of continuous functions*, Bull. Amer. Math. Soc. **55** (1949), 402–407.

[230] S. Myers, *Normed linear spaces of continuous functions*, Bull. Amer. Math. Soc. **56** (1950), 233–241.

[231] M. Nagasawa, *Isomorphism between commutative Banach algebras with an application to rings of analytic functions*, Kodai Math. Sem. Rep. **11** (1959), 182–188.

[232] B.Sz. Nagy and C. Foias, *Harmonic analysis of operators on Hilbert space*, North Holland Publ. Co., Amsterdam-London, 1070.

[233] H. Nakano, *Modulared semi-ordered linear spaces*, Maruzen, Tokyo, 1950.

[234] H. Nakano, *Modulared linear spaces*, Journ. Fac. Sci., Univ. Tokyo **I.6** (1951), 85–131.

[235] H. Nakano, *Topology and topological linear spaces*, Nihonbashi, Tokyo, 1951.

[236] A. Neyman, *Representation of L_p-norms and isometric embedding in L_p-spaces*, Israel J. Math. **48** (1984), 129–138.

[237] W. Novinger, *Linear isometries of subspaces of continuous functions*, Studia Math. **53** (1975), 273–276.

[238] W. Novinger and D. Oberlin, *Linear isometries of some normed spaces of analytic functions*, Canad. J. Math. **37** (1985), 62–76.

[239] D. O'Donovan and K. Davidson, *Isometric images of C^*-algebras*, Canad. Math. Bull. **27** (1984), 286–294.

[240] G. Okikiolu, *Differentiation of L^p-norms of linear expressions and representations of isometric operators in L^p-spaces*, Bulletin of Mathematics **5** (1982), 1–30.

[241] R. Paley, *A remarkable series of orthogonal functions I*, Proc. London Math. Soc. **34** (1932), 241–264.

[242] R. Paley, *Some theorems on abstract spaces*, Bull. Amer. Math. Soc. **42** (1936), 235–240.

[243] A. Paterson, *Isometries between B^*-algebras*, Proc. Amer. Math. Soc. **22** (1970), 570–572.

[244] A. Paterson and A. Sinclair, *Characterizations of isometries between C^*-algebras*, J. London Math. Soc. (2) **2** (1972), 755–761.

[245] V. Pathak, *Isometries of $C^{(n)}[0, 1]$*, Pacific J. Math. **94** (1981), 211–222.

[246] V. Pathak, *Linear isometries of spaces of absolutely continuous functions*, Canad. J. Math. **34** (1982), 298–306.

[247] V. Paulsen, *Completely bounded maps and dilations*, Pitman Research Notes in Math., Longman, London, 1986.

[248] G. Pedersen, *C^*-algebras and their automorphism groups*, Academic Press, 1979.

[249] A. Pelczynski, *On $C(S)$-subspaces of separable Banach spaces*, Studia Math. **31** (1968), 513–522.

[250] R. Phelps, *Extreme points of polar convex sets*, Proc. Amer. Math. Soc. **12** (1961), 291–296.

[251] R. Phelps, *Lectures on Choquet's theorem*, Princeton University Press, Princeton, 1966.

[252] R. Phelps, *Convex functions, monotone operators, and differentiability*, Lecture Notes in Mathematics, vol. 1364, Springer-Verlag, Berlin, 1989.

[253] G. Pisier, *An introduction to operator spaces*, Camb. Univ. Press, to appear.

[254] M. Plancherel, *Contribution à l'étude de la représentation d'une fonction arbitraire par des intégrales définies*, Rend. Circ. Math. Palermo **30** (1910), 289–335.

[255] A. Plotkin, *Isometric operators in spaces of summable analytic and harmonic functions*, Soviet Mat. Dokl. **10** (1969), 461–463.

[256] A. Plotkin, *On isometric operators on subspaces of L^p*, Soviet Mat. Dokl. **11** (1970), 981–983.

[257] A. Plotkin, *Continuation of L^p-isometries*, V.A.Steklova Akad. Nauk. SSSR **22** (1971), 103–129.

[258] P. Prased, *Isometries of $C^{(1)}([0,1]^m)$*, Tech. report, S.P. University, V.V. Nagar, India, 1990.

[259] C. Putnam and A. Wintner, *The orthogonal group in Hilbert space*, Amer. J. Math. **74** (1952), 52–78.

[260] M. Rais, *The unitary group preserving maps (the infinite dimensional case)*, Linear and Multilinear Algebra **20** (1987), 337–345.

[261] B. Randrianantoanina, *Isometric classification of norms in rearrangement-invariant function spaces*, Comment. Math. Univ. Carolinae **38** (1997), 73–90.

[262] M. Rao and Z. Ren, *Theory of Orlicz spaces*, Marcel Dekker, New York, 1991.

[263] N. Rao and A. Roy, *Linear isometries of some function spaces*, Pacific J. Math. **38** (1971), 177–192.

[264] T. Rao, *Isometries of $A_C(K)$*, Proc. Amer. Math. Soc. **85** (1982), 544–546.

[265] J. Ratz, *On isometries of generalized product spaces*, Siam J. Appl. Math. **18** (1970), 6–9.

[266] F. Riesz and B. Sz-Nagy, *Functional analysis*, Frederick Ungar, New York, 1955, Translated from 2nd French ed. by L. Boron.

[267] R. Roan, *Composition operators on the space of functions with H^p-derivative*, Houston J. Math. **4** (1978), 423–438.

[268] S. Rolewicz, *Metric linear spaces*, 2nd ed., Polish Scientific Publishers, Warzawa, 1984.

[269] H. Rosenthal, *Contractively complemented subspaces of Banach spaces with reverse monotone (transfinite) bases*, Longhorn Notes, The University of Texas Functional Analysis Seminar, 1984, pp. 1–14.

[270] H. Royden, *Real analysis*, third ed., Macmillan, New York, 1988.

[271] Z. Ruan, *Subspaces of C^*-algebras*, Ph.D. thesis, UCLA, 1987.

[272] W. Rudin, *Some theorems on bounded analytic functions*, Trans. Amer. Math. Soc. **78** (1955), 333–342.

[273] W. Rudin, *Real and complex analysis*, McGraw-Hill, New York, 1966.

[274] W. Rudin, *Functional analysis*, McGraw-Hill, New York, 1973.

[275] W. Rudin, *L^p-isometries and equimeasurability*, Indiana Univ. Math. J. **25** (1976), 215–228.

[276] W. Rudin, *Function theory in the unit ball of C^n*, Die Grundlehren der mathematischen Wissenschaften in Einzeldarstellungen, vol. 241, Springer-Verlag, New York, 1980.

[277] B. Russo, *Isometries of L^p-spaces associated with finite von Neumann algebras*, Bull. Amer. Math. Soc. **74** (1968), 228–232.

[278] B. Russo and H. Dye, *A note on unitary operators in C^*-algebras*, Duke Math. J. **33** (1966), 413–416.

[279] R. Schneider, *Isometries of $H^p(U^n)$*, Canad. J. Math. **25** (1973), 92–95.
[280] R. Schneider, *Unit preserving isometries are automorphisms in certain L^p*, Can. J. Math. **XXVII** (1975), 133–137.
[281] I. Schur, *Einige Bermerkungen zur determinanten theorie*, S.B. Preuss Akad. Wiss. Berlin **25** (1925), 454–463.
[282] Z. Semadeni, *Banach spaces of continuous functions*, Monografie Matematyczne, Warszawa, 1971.
[283] A. Shirayayev (ed.), *Selected works by A.N. Kolomogorov*, Kluwer Acad. Publishers, Dordrecht, Netherlands, 1992, Translated by G. Lindquist.
[284] J. Shohat and J. Tamarkin, *The problem of moments*, Math Surveys No. 1, Amer. Math. Soc., New York, 1943.
[285] G. Silov, *Ideals and subrings of the rings of continuous functions*, C. R. (Doklady) Acad. Sci. URSS **22** (1939), 7–10.
[286] I. Singer, *Sur l'extension des fonctionnelles lineaires*, Rev. Math. Pures. Appl. **1** (1956), 99–106.
[287] I. Singer, *Sur la meillure approximation des fonctions abstraites continues a valeurs dans unb espace de Banach*, Rev. Math. Pures. Appl. **2** (1957), 245–262.
[288] I. Singer, *Bases in Banach spaces I*, Springer-Verlag, Berlin, 1973.
[289] I. Singer, *Bases in Banach spaces II*, Springer-Verlag, Berlin, 1981.
[290] B. Solel, *Isometries of CSL algebras*, Trans. Amer. Math. Society **332** (1992), 595–606.
[291] M. Stanev, *Surjective isometries of Banach spaces of slowly growing holomorphic functions in ball*, Dokl. Akad. Nauk **334** (1994), 702–704.
[292] K. Stephenson, *Certain, integral equalities which imply equimeasurability of functions*, Can. J. Math. **XXX** (1977), 827–844.
[293] K. Stephenson, *Isometries of the Nevanlinna class*, Indiana Univ. Math. J. **26** (1977), 307–324.
[294] J. Stern, *Le groupe des isometries d'un espace de Banach*, Studia Math. **LXIV** (1979), 139–149.
[295] M. Stone, *Applications of the theory of boolean rings in topology*, Trans. Amer. Math. Soc. **41** (1937), 375–481.
[296] M. Stone, *A general theory of spectra, II*, Proc. Nat. Acad. Sci. U.S.A. **27** (1941), 83–87.
[297] W. Strobele, *On the representation of the extremal functionals on $C_0(T, X)$*, J. of Approx. Theory **10** (1974), 64–68.
[298] K. Sundaresan, *Some geometric properties of the unit cell in spaces C(X,B)*, Bull. de L'acad. Polonaise des Sciences **XIX** (1971), no. 11, 1007–1012.
[299] M. Takesaki, *Theory of operator algebras I*, Springer-Verlag, New York, 1979.
[300] K. Tam, *Isometries of certain function spaces*, Pacific J. Math. **31** (1969), 233–246.
[301] J. Tanaka, *On isometries of Hardy spaces on compact abelian groups*, Pacific J. Math. **95** (1981), 219–232.
[302] J. Tate, *On the relation between extremal points of convex sets and homomorphisms*, Comm. Pure Appl. Math. **4** (1951), 31–32.
[303] K. Taylor and W. Werner, *Differentiability of the norm in von Neumann algebras*, Proc. Amer. Math. Soc. **119** (1993), 475–480.
[304] E. Titchmarsh, *Introduction to the theory of Fourier integrals*, Oxford University Press, Oxford, 1937.
[305] A. Tulcea, *Ergodic properties of isometries in L^p spaces, $1 < p < \infty$*, Bull. Amer. Math. Soc. **70** (1964), 366–377.
[306] E. Vesentini, *Holomorphic families of holomorphic isometries*, Proceedings of the Conference on Complex Analysis, III, (College Park, Md.) (Berlin), Lecture Notes in Math., vol. 1277, Springer-Verlag, 1987, pp. 290–302.
[307] E. Vesentini, *Semigroups of holomorphic isometries*, Adv. in Math. **65** (1987), 272–306.

[308] E. Vesentini, *Isometries of Cartan domains of type one*, Atti Accad. Naz. Lincei Rend. Cl. Sci. Fis. Mat. Natur. Rend. Lincei(9) Mat. Appl. **2** (1991), 65–72.

[309] E. Vesentini, *Holomorphic isometries of Cartan domains of type iv*, Atti Accad. Naz. Lincei Rend. Cl. Sci. Fis. Mat. Natur. Rend.Lincei(9) Mat. Appl. **3** (1992), 287–294.

[310] E. Vesentini, *On the Banach-Stone theorem*, Advances in Math. **112** (1995), 135–146.

[311] E. Vesentini, *On the Banach-Stone theorem, II*, Rend. Addad. Naz. delle Sci. detta dei XL **XXII** (1998), 51–70.

[312] I. Vidav, *Eine metrische Kennzeichnung der selb-adjungierten operatoren*, Math. Z. **66** (1956), 121–128.

[313] J. von Neumann, *Allgemeine eigenwerttheorie hermitischer funktionaloperatoren*, Math. Ann. **102** (1929), 49–131.

[314] J. von Neumann, *Zur operatoren methode in der klassichen Mechanik*, Ann. of Math.(2) **33** (1932), 587–642.

[315] R. Wang, *Isometries of $C_0(X, \sigma)$ type spaces*, Kobe J. Math. **12** (1995), 31–43.

[316] R. Wang, *Isometries of $C_0^{(n)}(X)$*, Math. Japonica **44** (1996), 93–100.

[317] G. Watson, *General transforms*, Proc. London Math. Soc. **35** (1933), 156–199.

[318] N. Weaver, *Isometries of non-compact Lipshitz spaces*, Canad. Math. Bull. **38** (1995), 242–249.

[319] J. Wells and L. Williams, *Embeddings and extensions in analysis*, Springer-Verlag, Berlin-Heidelberg-New York, 1975.

[320] W. Werner, *Subdifferentiability and the noncommutative Banach-Stone theorem*, Proceedings of the Conference on Function Spaces, Edwardsville, Lecture Notes in Pure and Applied Mathematics, vol. 172, Marcel Dekker, 1995, pp. 377–386.

[321] H. Weyl, *Symmetry*, Princeton University Press, Princeton, New Jersey, 1952.

[322] R. Wobst, *Isometrien in metrischen vektorraumen*, Studia Math. **54** (1975), 41–54.

[323] H. Wold, *A study in the analysis of stationary time series*, Almquist and Wiksell, Uppsala, 1938.

[324] J. Wright and M. Youngson, *Isometries of Jordan algebras*, J. London Math. Soc. **17** (1978), 339–344.

[325] F. Yeadon, *Isometries of non-commutative L^p-spaces*, Math. Proc. Cambridge Philos. Soc. **90** (1981), 41–50.

[326] M. Zaidenberg, *Groups of isometries on Orlicz spaces*, Soviet Math. Dokl. **17** (1976), 432–437.

[327] M. Zaidenberg, *On isometric classification of symmetric spaces*, Soviet Math. Dokl. **18** (1977), 636–640.

[328] M. Zaidenberg, *Special representations of isometries of functional spaces*, Investigations on the theory of functions of several real variables (Y. Brudnyi, ed.), University of Yaroslavl', 1980, pp. 84–91 (Russian).

[329] M. Zaidenberg, *A representation of isometries of function spaces*, Institute Fourier (Grenoble) **305** (1995), 1–7.

Index